Managing Complex Construction Projects

A Systems Approach

Best Practices and Advances in Program Management Series

Series Editor
Ginger Levin

Managing Complex Construction Projects

A Systems Approach

John K. Briesemeister

CRC Press
Taylor & Francis Group
Boca Raton London New York

CRC Press is an imprint of the
Taylor & Francis Group, an **informa** business

AN AUERBACH BOOK

CRC Press
Taylor & Francis Group
6000 Broken Sound Parkway NW, Suite 300
Boca Raton, FL 33487-2742

First issued in paperback 2022

ISBN 13: 978-1-03-247628-5 (pbk)
ISBN 13: 978-1-4987-8311-8 (hbk)
ISBN 13: 978-0-429-48668-5 (ebk)

DOI: 10.1201/9780429486685

Visit the Taylor & Francis Web site at
http://www.taylorandfrancis.com

and the CRC Press Web site at
http://www.crcpress.com

Dedication

This book is dedicated to all of the men and women working in the construction industry throughout the world in various positions who, through their knowledge and sweat, are building dreams "one day at a time."

Contents

Foreword

The Construction Industry is known for many programs and projects being behind schedule or over budget or experiencing both of these issues given the complexity of the work to be done. John Briesemeister, winner of the 2016 David S. Barrie award from the Project Management Institute, has managed construction projects on various sites around the world and continues to work in this industry. These projects are known for their challenges and complexity as many are in remote areas of the world. His expertise in avoiding cost and schedule issues is extraordinary...He has combined his on-site experience, along with the knowledge gained by his third Master's degree in Project Management and a previous Master's degree in Industrial Engineering, to develop the field-based Construction Management approach discussed in this book...He has used this approach on various large, complex construction projects around the world...This book is a "must-read" for anyone working in the construction field or considering it. It also is one that anyone interested in this topic will find interesting and useful...I hope you enjoy reading it and learning from it as much as I did.

— Dr. Ginger Levin, PMP, PgMP, OPM3

Preface

In this toolbox or book, as some would call it, the reader will find three systems that, used together, can be used to effectively manage the complex construction of large, complex projects. Each of these systems should be viewed as a tool that was designed with over twenty years of formal education and forged in the fires of more than thirty years in the field of onsite construction.

If a person is tired of welding piping in the trenches or bolting up steel structures, there is the Work Management System that will help him or her move out of that trench or off of that structure into a career as a labor superintendent, which could potentially lead this person to managing his or her own construction company someday.

If a person would like to know more about developing a Quality Management program for the construction work onsite, or move from a quality inspector into being a QA/QC Manager, there is the Quality Management System.

Finally, if a person wants to move from college into the exciting field of construction management, with all of its unique challenges, and work with some amazing people, there is the Project Management System that will start him or her on this journey. If this same person wants to see the big picture and effectively paint that big picture, then he or she should learn all three systems in this book and how these systems, when combined, can transform an empty piece of land into a bright new future for millions of people.

Acknowledgments

I would like to thank Professor Ginger Levin, who inspired me to write this book and worked with me on its development.

I would also like to thank John Wyzalek (Taylor & Francis Group) for providing this opportunity and for his support in the publishing of this book.

In addition, I would like to thank Theron R. Shreve, Director, and Marje Pollack, copy editor and typesetter (DerryField Publishing Services)—who worked tirelessly and with great precision—for their collaboration and dedication in the production of this book.

I also give thanks to the Lord above, who gave me the knowledge and wisdom that made this endeavor possible.

About the Author

The author, who is a Vietnam Veteran and a licensed Project Management Professional since 2007, began his construction experience as a Site Senior Mechanical Engineer on a large nuclear construction project upon graduation from the University of Minnesota in June 1981.

He has continued working in the power industry as an Engineer, Construction Manager, Site Manager, and Project Manager since the mid-1980s, and he expanded his academic education to include an MBA in 1994, a Master's degree in Industrial Engineering (MSIE) in 2003, and a Master's degree in Project Management (MSPM) in 2015.

This author has successfully worked in the field on various large, complex construction projects for the past thirty years and last year, 2016, was awarded the 2016 Donald S. Barrie Award in Construction Management at the PMI® Global Congress 2016—North America in San Diego, California, from the Project Management Institute Educational Foundation (PMIEF) for a paper he wrote and submitted, which was titled, "Construction Execution Plan Alignment for Successful Construction Projects."

Chapter 1

Introduction

1.1 Introduction

If we take the standard multi-story apartment building and add another four buildings, a substation for the electric power needs, solar heating on all of the roofs, a sports complex for the inhabitants that includes a swimming pool, and, finally, a special water filtration plant for this complex, then most Program, Project, and Construction Managers would agree this is now a complex project.

To inexperienced Program, Project, or Construction Managers, a complex project seems to be a labyrinth with many hidden dangers and is difficult to manage. It is the intent of this book to provide best practices and insight for Program, Project, or Construction Managers so they can not only identity these hidden dangers but also effectively manage the construction process to either mitigate or eliminate these risks. The approach that will be presented in this book is based upon three systems, as shown in Figure 1-1.

These three systems are as follows:

1. Project Management System
2. Work Management System
3. Quality Management System

The interface areas shown in Figure 1-1 are also important and must be effectively managed because of the interdependencies among these three systems. The problem with complex programs and projects is that many Construction, Program, or Project Managers are only equipped with a knowledge of project management, which is only one of the systems.

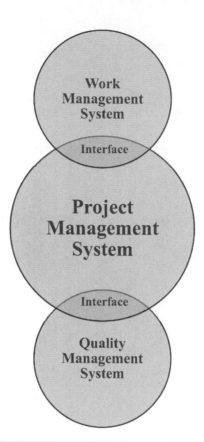

Figure 1-1   Construction Three-System Relationship Diagram

A system for construction is a collection of many processes effectively work-ing together for the production of a specific product or deliverable, which is usually defined in the program or project's contract. This system has a series of specific inputs and outputs, which are what the customer expects from the com-pany or companies performing the work. This will be the approach that will be taken in this book so that the Construction, Program, or Project Manager managing the construction work can use a checklist when he or she first arrives onsite. This checklist will comprise the inputs and outputs for each system that will be clarified and discussed within this book.

In order for this concept to be clear, we will briefly look at a similar project that most people are familiar with—the education of a child. This project pri-marily consists of two systems, as shown in Figure 1-2.

The information for this project, along with the inputs and outputs for each system, are shown in Table 1-1.

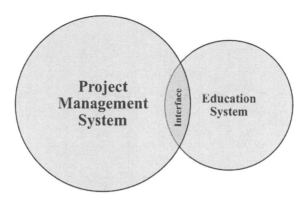

Figure 1-2   Education of a Child Two-System Relationship Diagram

If we examine Table 1-1, we see that both systems have the same output and require both time and money, which is one of their interface areas. The interface area shows that in order for the parent to monitor the progress of this project,

Table 1-1 System Analysis for the Education of a Child Project

Name of Project	Education of a Child		
Duration	12 Years		
Budget*	$148, 812 ($12,401 per year* for 12 Years)		
System	Inputs	Outputs	Remarks
Project Management	Child	Educated Child	
	Capital		
	Time		
	Nutrition		
	Training		
Education	Capital	Educated Children	
	Qualified and Certified Teachers		
	Training Materials		
	Buildings (Schools)		
	Transportation		
	Time		
	Children		

* 2011–2012 figures provided by the National Center for Education Statistics (http://nces.ed.gov/fastfacts/display.asp?id=66).

he or she needs to have progress reports, which are supplied by the Education System at specified intervals. If the progress of the project does not achieve the desired output within 12 years, then the parent or Project Manager may increase the training provided, which will incur additional cost for the project.

It may surprise some parents that they have been practicing project management, but this simplistic example clearly shows how educating one's child has its own degree of complexity, from a project perspective.

This book is written to provide a "nuts and bolts" approach for the Program, Project, or Construction Manager once he or she arrives onsite at a complex project. How many of us have either experienced or observed the following at the construction site?

- Contractor personnel are working in areas that are not on the critical path.
- In a meeting with the contractors, the term "baseline" is considered to have more to do with baseball than with schedules.
- The main contractor's Site Project Manager tells you that there is no "baseline schedule," but he or she knows that the project is on track.
- Contractor personnel are walking off the project due to pay issues with their current employer.

This list can go on indefinitely, but all of this points to very serious signs of trouble for the project. The system approach in this book can fix this, if applied early—at the first signs of trouble or during the project initiation phase. The topics we will follow in this book are the following:

Chapter 2 – Project Management System, Part I
Chapter 3 – Project Management System, Part II
Chapter 4 – Project Management System, Part III
Chapter 5 – Work Management System
Chapter 6 – Quality System
Chapter 7 – Bringing It All Together
Chapter 8 – Lessons Learned from the Field

In Chapters 2 through 6, the three systems will be defined, analyzed in detail, and developed so that the reader will not only understand each system but will also be able to apply each of them effectively in the field.

In Chapter 7, we will take the complex project briefly discussed at the beginning of Chapter 1—Introduction and walk the reader through the process of applying each system to effectively manage this complex project and make it a success for both the EPC Company and the Owner.

In Chapter 8, the author will provide many valuable "lessons learned" for the reader, which are based upon the author's more than thirty years of working on various construction sites around the world. These "lessons learned" will not only provide valuable information to the reader for future projects, but they will also make the reader more effective as a Program, Project, or Construction Manager when managing a complex project onsite.

This book will also open up a new world of construction terminology and words for the reader, such as "Approved for Construction (AFC)," "Field Action Requests," "Site Survey," etc., which can be found at the end of this book in the Glossary section. This Glossary includes a brief description of each term to facilitate a full understanding not only of the word or term but also how it applies to the construction process and which systems it applies to.

1.2 Concluding Remarks

Since the Project Management System is very large and has its start at the time of project initiation, it will be discussed in detail in the next three chapters. These three sections are designed to bring the reader from the time of project initiation through the project planning phase and into the project execution phase, which is why the Project Management System is the backbone of the complete process of managing large, complex construction projects.

Chapter 2

Project Management System: Part I

2.1 Project Management System

The Project Management System (PMS) is the most critical system in the effective management of a complex construction project. However, without vital inputs from the Work Management and Quality Systems, it will become ineffective because these two systems provide constant feedback to the Project Management System during the construction process so that the Program, Project, or Construction Manager makes the right decisions at the right time. It is critical that the reader fully understands the interdependency of these systems so that he or she can take immediate action onsite, if necessary, to ensure that the feedback channels into the PMS are providing the consistent and accurate inputs that this critical system requires over the life of the project. In this chapter, we will discuss how the PMS is developed through the bidding process for a large, complex construction project. After the bid is submitted and accepted, we will then discuss how the various components of the PMS are developed further and how governance is established for the project through alignment within the PMS.

2.2 Initiation Phase

The PMS, unlike the next two systems (which are site related), has its beginnings in the Initiation Phase of the project, but it will change over the next four phases of the project, especially during the Execution Phase when construction starts and continues onsite. It has its start with the Project Bid meeting, which is usually conducted by the customer of the project through the solicitation of bids for a specific project workscope. This one of the most critical points for the project, for the following reasons:

1. An Engineering, Procurement, and Construction (EPC) Company needs to be fully prepared for this meeting, and its representatives should not just present the bid and conclude the meeting since there will be other EPC Companies at the bid meeting competing for the same project workscope. The EPC Company should already have a list of questions derived from their evaluation of this bid document, which can be just a few pages in length or a stack of volumes, and they should bring this list to this meeting. At this meeting, if there are any questions about the bid document, the EPC Company should bring them to the customer's attention during the discussion.

2. The timetable established by the customer of the project for bid submittal should be discussed and evaluated to ensure that the EPC Company's representatives will have sufficient time to present a concise and accurate bid. The budget developed by the Program or Project Manager must accurately reflect all of the costs required to perform the workscope within the contract, starting with the Bid Meeting.

3. At this meeting, the EPC Company's representative should establish with the customer a date for a Site Walkdown Meeting, which is usually attended by all EPC Companies bidding on the work. This should be done after the EPC Company has sufficient time to evaluate the bid and develop a list of items to be checked off during this walkdown. If this is not done, this failure will lead to many future problems onsite during construction, which will be much harder to resolve at that time rather than in the bidding process and could lead to a cost or schedule impact for the project.

4. After this Site Walkdown Meeting is complete and the results evaluated, the next step is to contact the customer and request a Bid Clarification Meeting, which must be attended by all EPC Companies bidding on the work. It is unethical and illegal to simply have a meeting with the customer to discuss areas of the bid where a company has questions because this would give that company an unfair advantage over the other bidders and could lead to the company being banned from any further work with the customer or in the industry.

2.2.1 Project Bid Evaluation

After the meetings discussed above have been completed, and the Program or Project Manager has confirmed with management that the EPC Company is confident to start work on the compilation of a bid for this project, a schedule should be established immediately for all parties that matches the customer's bid submittal timetable. Many EPC Companies tend to bring the Program or Project Manager into a project after the bidding process, but the best practice is for the Program or Project Manager to be involved from the beginning of the bid process and follow the project all the way through until it is completed and closed. The organizational chart in Figure 2-1 is just a small example of how many parties can be involved in this compilation process, which is usually presented and reviewed by the CEO of the company and its Board of Directors, upon completion.

Bid Compilation Organization

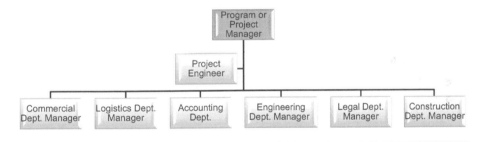

Figure 2-1   Organization Structure for Bid Compilation

The steps to be followed for a proper Bid Compilation should be as follows:

Step 1 – Project Work Breakdown Structure (WBS)

The first and very critical step of this process, which must be completed before the Engineering, Logistics, and Construction Departments can get started, is the development of a Level 4 Work Breakdown Structure (WBS) that not only includes all of the work in the contractual bid but must also exactly mirror the way it will be performed onsite through the work packages at this level. In a complex construction project, there are many Engineering discipline interface problems, such as mechanical versus electrical, civil versus mechanical, etc., and this is where they need to be identified, not onsite during construction. An example of this type of Level 4 breakdown is shown in Table 2-1.

Table 2-1 Example of a Construction WBS Level 4

WBS No.	Engineering Discipline	WBS Level	Activity Description
1.0	Civil	1	Civil
1.1	Civil	2	General Site Civil Work
1.1.01	Civil	3	Site Preparation
1.1.01.01	Civil	4	Site Survey Work

The Level 4 breakdown is the point at which Construction Work Packages are produced because, at this level, the Construction Manager and the Construction Department can evaluate the following:

- The work that is required to be performed
- The equipment that will be required for this work
- The personnel that will be required along with supervision for this work
- The type of work entailed, because if it is a specialty type of work, a separate contract may have to be produced for the vendor

For some projects, the WBS level at which this work is completed can vary, but the emphasis should always be on the level at which the work packages will be produced. This construction work evaluation should be performed with input from the Engineering Department that is producing the design and respective drawings for the contractual deliverables that the project must produce. If a WBS is presented separately from each department without this interaction, then there is a high risk of scope creep to the project once construction starts, which can have severe consequences to the project's budget and schedule. After this evaluation is completed, then the WBS should be reviewed, and all changes recorded.

Step 2 – Project Budget

After the first step is completed, the next step is to evaluate the WBS and allocate human resources, personnel, and equipment required for each work package at Levels 4 or 5 of this WBS. If a Program Manager is utilized, then the Project Managers assigned to various projects within the overall construction project must also be included in the resource allocation and cost for the project. After this allocation is completed, then the total number of person-hours can be compiled along with a total for all material and equipment costs, which should provide the Program Manager or Project and Construction Managers with the budget for this project. If the company has done similar projects in the past, it is good idea to evaluate this latest budget and compare it what was used in the past, if there are differences, then evaluate those differences to ensure that this budget is a truly accurate figure of what it will cost to complete this project. This project budget cost will be then increased by management for a specific margin of profit before it is submitted in the final bid. A sample of how this process works is shown in Table 2-2.

Table 2-2 Example of Developing the Budget for Just One Activity

WBS No.	Description	Quantity/ Type	Rate $/Hr.	Cost for 40 Hrs./Wk.	Total	Bid Price (5%) Margin
1.1.01.01	Site Survey Work	4 /Surveyors	$80	$12,800	$12,800	$13,440

If the duration of this work package, for example, was more than one week, then the total amount would be multiplied by the actual number of weeks anticipated for the completion of this activity. Most companies usually add an additional percentage to this final figure to provide a margin to cover any fluctuations during operation, which can be as high as 5%, in some cases.

If the company feels that there is a certain amount of unidentifiable risk that is due to this project's location, type of work, etc., then executives may add an extra amount of money to the final bid price to cover it. This is called a "contingency fund," which is not part of the project's budget and is usually strictly controlled by the company's senior management.

Hidden Costs

A large, complex construction project has a much higher risk due its complex nature and the length of the project. Therefore, the Program or Project Manager, with the help of the Construction Manager, should attempt to identify as many "hidden costs" as possible at this stage of the process, prior to the submittal of a bid for the project. Some of these "hidden costs" are as follows:

A. **Soil Improvement** – The customer may provide a geophysical survey of the area where the project's construction work will be performed, but it may not be complete. For example, if the project will involve work both on land and offshore—that is, a pier, a beach, etc.—then this survey should include both areas, onshore and offshore, which in some situations is overlooked. Soil improvement usually involves many expensive ways to increase the load-bearing capacity of soil because what exists onsite is typically insufficient to support the structures being built. This may include activities such as pile driving, stone column replacement, grout injection, etc., which are costly and can significantly delay the project. If this situation is found after the project is started onsite, the customer is usually reluctant to accept any change orders from the Program Manager or Project Manager for this work because the customer feels that the company should have raised this concern at the time of bid submittal.

B. **Labor Mobilization** – In the United States, this is usually not that significant because the labor, union or non-union, is locally sourced, and the process to get site permits for these workers is fairly simple. However, this is not the case in the international arena, where many countries have lengthy processes that must be followed for obtaining a working visa for each member of the workforce as well as the restrictions on the number of foreign workers or professionals that can be brought in for the project. In addition to these requirements, some countries also stipulate that a certain percentage of this workforce must be from that country, which forces the Program Manager or Project Manager to deal with local labor brokers, and the local people may not have the required skills required for the project. If this occurs, then the cost of training these workers must also be added to the other costs, to comply with this local government requirement. This situation may not seem to have such a great impact on the project, but if the large, complex project will require a workforce of 5,000 people at its peak, the cost and delay to the project can be huge. Most customers do not consider this to be a recoverable cost for the EPC Contractor because it is felt that the contractor should have known about this situation at the time it submitted its bid for the project.

C. **Infrastructure** – A project needs to have a solid infrastructure in order for the construction work to be performed effectively, and communication with all of the project stakeholders should be maintained over the life of the project. If the construction site for the project is in a developed area, this is not an issue, in most cases. However, if this site is a "green field" project, which means an area that has never been cleared of the local vegetation, and, in many cases, does not have roads, then this is something that needs to be addressed during the bid-submission process. It can be addressed by asking the following questions:

- Do the surrounding roads have the capacity to support the heavy items (which can weigh up to 100 tons or more) that have to be brought to the construction site? If not, who will upgrade these roads, and when will this upgrade be completed?
- What is available for internet support? If a satellite system is going to be required, how much will it cost for the amount of bandwidth that will be required for the transfer of large drawing files, what permits are required, and when will it be readily available?
- What types of site offices are required, and who will be responsible for their support in terms of water, sewage, electricity, etc.? This needs to be one of the first items established onsite so that the engineers and Construction Manager can immediately start the work for the project, which must include all of the required support services mentioned.
- If it is a project that will be an addition to an existing complex, then what are the interface points and how will they be managed? For example, will the EPC Contractor have access to potable water or drinking water for the construction site from this complex, and will there be an additional cost for the water? This may extend to other systems, such as firefighting systems, data-transmission systems, electrical systems, etc., so it is an important point to raise in the bidding process.
- Usually the customer handles site permitting, which should not be an issue. However, this process needs to be clarified at the time of bid evaluation. In many countries outside the United States, this process usually requires the agreement of the local municipality to give permission for the land the project will need, and this usually comes with various conditions attached. Those various conditions can dramatically impact the project and are an external risk that the Program Manager or Project Manager has to identify and closely monitor over the life of the project.

These are just some of the major hidden costs that will have a major impact upon the project since they occur at the time the Program Manager, Project Manager or Construction Manager is trying to get the construction work started onsite. If the Program Manager or Project Manager feels that all of the information to quantify or identify these hidden costs, then it is necessary to request a bid clarification with the customer to close this issue prior to bid submittal. After these hidden costs have been evaluated, the EPC Company's management may decide to add either additional money to the project to mitigate the financial risk to the project or additional time to the duration of the activities that may be impacted by these hidden costs.

Step 3 – Project Schedule

The third step is to now transform the WBS established for this project into Levels 4, 5, or another appropriate Level critical path schedule as specified by the customer for bid submittal, provided it shows each work package along with its duration. The Program or Project Manager, with the assistance of the project team, must ensure that this schedule meets the milestone dates specified in the customer's request for a bid because for most complex projects, payment to the EPC Company only occurs at the time a milestone is completed onsite. The EPC Company also has to ensure that it provides a schedule using the scheduling software specified in the customer's request for a bid because failure to do so will usually result in a rejection of the bid. The other important item that the Program or Project Manager must ensure is that the activity start and end dates established in the critical path schedule are achievable by all parties. For example, if the schedule states that the foundation work for the main building will start on August 14th, then both Engineering and Logistics must also agree that they can meet this date with the right "Approved for Construction (AFC)" drawings and materials.

Project Risk

In the compilation of a schedule for a complex construction project, it is essential that, for each activity, there are specific dates established, which is shown in Table 2-3.

Table 2-3 Example of Early Start and Finish Dates and Late Start and Finish Dates for an Activity

Activity	Early Start Date	Early Finish Date	Late Start Date	Late Finish Date
Activity A				
Site Survey Work (WBS No. 1.1.01.01)	9/04/18	11/25/18	09/10/18	01/02/19

In Table 2-3, the site survey work will normally take about three months to complete, but if the site excavation work (which is a project milestone) must start on January 1, 2019, then the Program or Project Manager now knows that they cannot permit this activity to slide out to September 10th, the Late Start Date, because the survey work will not be completed in time to support the start of excavation. In order for the Program, Project, or Construction Manager to determine the risk of this project being successfully completed on time and meet the specified milestone dates, a simulation of this schedule must be performed. The critical path for a project is always the path that takes the longest time to complete one activity. This activity has a fixed normal duration that sits inside an envelope of best and worst durations with the first one being the shorter duration and the second one being the longer duration, which are used in this simulation to establish an early and late completion. The first part of this simulation will be evaluating the project's completion and milestone dates using just the Early Start and Early Finish dates, which is called a "Forward Pass." The second part of this simulation will compromise evaluating the project completion and milestone dates using just the Late Start and Late Finish dates, which is called a "Backward Pass." This sounds complicated, but there are a number of scheduling risk software packages on the market, such as Palisade, Intaver Institute, Safran, etc., that will do this with a schedule using the Monte Carlo simulation without too much effort on the part of the responsible individual. The results of the simulation will now provide the Program or Project Manager with a time envelope for the project. This time envelope will provide the earliest expected completion date and the latest expected project completion date for the project, which can now be used to determine if the project, as currently planned and scheduled, will successfully meet the contractual dates specified. For example, this simulation may show that if the project starts two weeks earlier, then it will be completed two weeks earlier than planned, but if it starts just one week late, then the completion date will be three weeks later than planned. This compounding relationship is important to understand because it clearly shows that if the project slides just one week, then the Program Manager or Project Manager can expect the contractual completion date will be impacted by much more than one week, which may be not be recoverable. The compounding arises from the fact that in a large, complex construction project there are usually many other activities that are within thirty days or less from being on the project's critical path, and any delay brings them closer, which can create multiple critical paths for the project. This translates into a much higher risk for the project to fail because on most complex construction projects, it is very easy for a critical path activity, such as foundation excavation, to slide one week because of inclement weather at the project site.

After schedule risk is determined, the Program or Project Manager needs to evaluate if there is any financial risk to the project's budget. This evaluation looks at both internal and external risks, especially if it is a project located in a foreign county. In the United States, a labor strike on a complex construction project can have a severe financial impact upon a project, and this risk needs to be identified prior to bid submittal because the EPC Company's management should not only know this possibility but also be able to determine what impact it will have on the company's balance sheet and its stockholders. If this project is in a country outside the United States, local unrest may preclude access to the site by the foreign EPC Company and require government intervention, which takes time for the customer to resolve. This risk needs to be carefully evaluated, along with its impact on the project and the EPC Company. This risk evaluation should also involve input from the project team members because each person, based upon his or her experience, may have a different view as to what poses a schedule or financial risk to the project.

2.2.2 Project Presentation

After the bid has been evaluated and prepared for submission to the customer for review and possible acceptance, the final hurdle prior to submission is approval from either the EPC Company's management or Board of Directors because there is an internal cost to the company for preparing a bid, which can be substantial, depending upon the size of the project. This approval is received only after the Program or Project Manager has successfully presented the project to senior management of the company or its Board of Directors, which includes successfully answering all of the questions that are brought up during this presentation. This means that the person doing the presentation has to be fully prepared and well skilled in the art of conducting effective presentations. The items that should be included in this site presentation are the following:

1. **Project Description** – This should be a brief overview of the project and include a map of the area where the work will be executed. The Program or Project Manager should also show how this project will benefit the company and meets its vision.
2. **Project Budget** – This should be a concise and clear breakdown by category of all costs associated with the budget for this project. The numbers presented are important, but it also shows the audience that the Program or Project Manager has covered all of the bases, which is even more important. During the presentation, this section is where the change order process for this project should be presented, along with how it will be managed onsite during the construction process.
3. **Project Schedule** – This should be a high-level view of the project's schedule, which includes all of the planned contractual milestone dates and the planned completion date of the project. This schedule is important because it will be the basis for the project's cash flow over the life cycle of this project, especially if the contract includes milestone payments.
4. **Project Risk** – This should be a brief, but concise, list of the risks that this project currently faces at this time, which will change once the

project's construction work starts onsite. A graphical risk matrix should be presented at this time so that the senior management knows which risks have the highest impact upon the project, as shown in Figure 2-2.

Probability					
0.90	0.90	1.80	2.70	3.60	4.50
0.70	0.70	1.40	2.10	2.80	3.50
0.50	0.50	1.00	1.50	2.00	2.50
0.30	0.30	0.60	0.90	1.20	1.50
0.01	0.10	0.20	0.30	0.40	0.50
Impact -------	1	2	3	4	5

Low ➡ High

Note: Threat is established by Multiplying Probability of Occurrence by Impact

Figure 2-2   Project Risk Matrix Example

The numbers in Figure 2-2 are just for the discussion in this book because the numbers for impact, along with probability, typically depend upon the EPC Company and its standards for risk calculation. A schedule for the risk reviews at the end of each project phase should also be presented so that senior management, who are stakeholders for this project, have confidence that this area of the project will be effectively managed.

5. **Project Organization** – This is not only an organizational chart with names. It also includes a brief discussion of the responsibilities that each position will have for this project. At this time, the Program or Project Manager should also mention the location of each person, office, or site so that the company's management understands how the communication process will flow during the project.

6. **Project Communication** – This should be a brief overview of how the company's accounting department will receive its required financial reports for this project, and how the company's senior management will be regularly informed about the project's progress, once it starts. This is a good area for the Program or Project Manager to receive feedback from the company's senior management on what their communication preferences for this project are and how often they want to receive updates.

7. **Project Health, Safety, and Environmental Plan (HSE Plan)** – HSE issues are becoming more critical to companies, and this plan plays an important role in the effective management of the construction process onsite. HSE issues can expose a company not only to litigation but also

to unwanted public scrutiny in the event that something goes terribly wrong. For example, the discharge of a Hazardous Chemical from a construction project into the local waterways can quickly ruin the reputation of a company, possibly impact its stock price, and jeopardize its future business opportunities in the industry.

8. **Project Quality Plan** – This should comprise a brief overview of how the quality of work performed onsite along with the respective business processes will be controlled to meet the specified quality standards set by both the contract and the EPC Company. It should also include the quality audits that will be performed over the life of the project, what they will include, and how often they will be conducted, along with who will be performing them.

During this meeting, the Program or Project Manager should have someone recording all of the questions raised, to assist in developing a list of action items required to be closed prior to submission of the bid. At the end of this meeting, the Program or Project Manager should not only leave time for questions but at the same time should receive a firm commitment from the company's senior management or Board of Directors that they approve this project and will solicit their support for its success.

2.2.3 Project Bid Submittal

After the Program or Project Manager receives approval for the project, he or she can proceed with submittal of the bid to the customer for its evaluation and possible approval. If the company's management does not feel that they can accept certain parts of the bid as presented, then the company must include a list of exceptions along with its bid, but the company must also realize that too many exceptions may lead to a rejection of the bid.

2.3 Planning Phase

After the Initiation Phase is completed, assuming that the bid was accepted, a phase gate review should be conducted by the Program or Project Manager to capture "lessons learned" for future bidding processes and determine if some of these are applicable to the Planning Phase. In this Planning Phase, some of the basic inputs for the PMS established in the bidding process will be developed further and combined with additional inputs so that the PMS will be ready for implementation as this project moves into the Execution Phase onsite.

A graphical example that lists the inputs and outputs for the PMS is shown in Figure 2-3.

Project Management System Inputs and Outputs

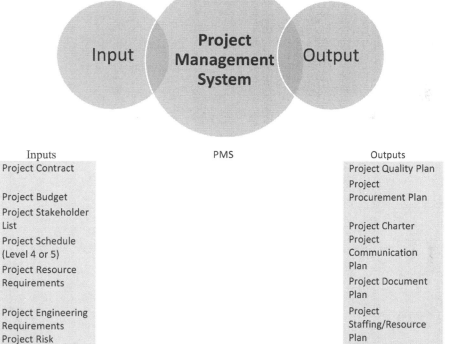

Inputs	PMS	Outputs
Project Contract		Project Quality Plan
		Project Procurement Plan
Project Budget		
Project Stakeholder List		Project Charter
Project Schedule (Level 4 or 5)		Project Communication Plan
Project Resource Requirements		Project Document Plan
Project Engineering Requirements		Project Staffing/Resource Plan
Project Risk Register		Project Logistics Plan
Project HSE Requirements		Project Risk Management Plan
		Project Financial Register
		Project Management Plan
		Project HSE Plan
		Project Construction Execution Plan
		Project Engineering Plan

Figure 2-3   Inputs and Outputs of the Project Management System (PMS)

**Table 2-4 Project Management System Inputs
and Their Relationship to Outputs**

Input Name	Information Provided	Output Affected
Project Contract	Scope of work Total cost of project Project location Milestone payment schedule Rights and responsibilities of all parties involved with the project Rules and regulations regarding legal jurisdiction over this project Penalty clause and what initiates it Litigation process Technical requirements Programs required prior to start of work Permit-to-Work (PTW) Authorization Process Reporting requirements. Customs clearance location and requirements Project deliverables	Project Charter Project Management Plan Project Financial Register Project HSE Plan Project Quality Plan PROJECT CONSTRUCTION EXECUTION PLAN Project Logistics Plan Project Communications Plan Project Engineering Plan Project Procurement Plan
Project Budget	Cost breakdown for the project Milestone payments Final planned cost at time of project completion	Project Financial Register Project Communications Plan Project Document Plan
Project Stakeholder List	Stakeholder name and company Stakeholder communication requirements Stakeholder location	Project Management Plan Project Communications Plan
Project Schedule (Level 4, 5, or Work Package)	Level requirements for reporting Critical path for the project Interface points with other subcontractors or projects	Project Management Plan Project Communications Plan Project Procurement Plan Project Engineering Plan
Project Resource Requirements	Personnel requirements over the life of the project Skilled labor requirements Professional personnel requirements over the life of the project Subcontractor requirements	Project Management Plan Project Communications Plan Project Staffing/Resource Plan Project Procurement Plan
Project Engineering Requirements	Quantity of Engineering drawings and documents for the project Identification of parties responsible for Engineering and location Process for drawing approval Drawings required for each activity	Project Management Plan Project Communications Plan Project Engineering Plan Project Quality Plan
Project Risk Register	External threats to the project Internal threats to the project Opportunities for the project Activities with the highest risk	Project Management Plan Project Risk Management Plan Project HSE Plan
Project HSE Requirements	HSE rules and regulations to be followed on site (local and international) Day-to-day requirements for safety program HSE threats locally and internally HSE management requirements	Project Risk Management Plan Project HSE Plan Project Management Plan Project Staffing/Resource Plan

It is imperative that the Program or Project Manager understand how the inputs are related to not only the outputs for this system but exactly which outputs and what their requirements are for them to be effective. In Table 2-4, each input is examined, and the corresponding output or outputs requiring their information are shown.

The effective and accurate development of the PMS's outputs will require the effort of the original group assembled for the bidding process combined with that of the Site Construction Manager, the Site Project Team, and other Subject Matter Experts (SMEs), as required. The Program or Project Manager has to ensure that all departments of the company, such as Engineering, Accounting, Procurement, HSE, etc., are in alignment with the project's schedule and budget so that project support is from the "top down" not from the "bottom up."

2.3.1 Outputs

The outputs of the PMS are the basic building blocks for the system and must all work together effectively in order for the project to be a success. This requires the Program, Project, and Construction Managers to fully understand each one of the outputs and their interactions within the PMS, which will now be discussed.

A. **Project Charter**

After the contract is received from the customer, the Project Sponsor within the EPC Company will put to paper the Project Charter, which will give the Program or Project Manager legitimate authority to manage the complex construction project. The EPC Project Sponsor is not only the champion for the project but also works directly for the senior management company to coordinate the company's various departments for provision of the corporate support that the project will need to be successful. It should also be noted that on the customer's side there should also be a Project Sponsor who works directly for the senior management of the company and has the responsibility of overseeing the entire project from its inception until all of the deliverables have been received. It is critical that within the Project Charter, the EPC Project Sponsor not only provides a vision for the project but, at the same time, also ensures that this vision is aligned with the company's vision. After this is done, then the Project Sponsor should provide a high-level view of this project from the standpoint of its WBS, its schedule (Level 2), its budget, its purpose and deliverables, its part in the company's business plan, and how its success will lead to success for the company. The next step is to introduce the Program or Project Manager who was involved with bidding the project to the customer's senior management so that they feel comfortable with the person selected by the Project Sponsor. The standard for most construction projects is to only assign a Project Manager, but when it comes to a very large, multibillion-dollar complex construction project, an experienced Program Manager is the better choice because he or she will be managing multiple projects, which will be discussed further in this book. The last step is to fill out the list of personnel that will be required to review and approve this Project Charter, which includes the Project Sponsor, so that it can be recorded in the company's records prior to the start of work onsite.

B. **Project Management Plan**

The Project or Program Manager usually has a brief Project Management Plan (PMP) that he or she developed during the bidding process because most companies' senior management requires this plan in order for them to provide approval for the project. After the Project Charter is issued, then the Project Management Plan only needs to be updated with this information with brief synopsis of the critical points included in the Project Charter. The other critical item is alignment of the project's goals with that of the EPC Company. This plan must provide the methodology that the project or program will follow over the life of the project to ensure that this critical alignment is maintained at all times. This is a challenge on a complex construction project because of the large number of subcontractors performing the work, but if the Program or Project Manager establishes the necessary controls before work starts, then maintaining this alignment is much easier. The other areas that the Project or Program Manager must ensure are in the PMP are the following:

1. **Project Scope Management** – This area should provide a Level 5 WBS and a corresponding detailed work package for each activity at this level, which was developed during the project's bidding process, to clearly show the full breadth of the project's work scope and associated work packages. The next step after this is to define how the Project or Program Manager plans to effectively manage this work to prevent scope creep and what Key Performance Indicators (KPIs) will be employed for consistently monitoring the progress of these activities as they are performed by the various subcontractors. The critical item in this step is to determine the benchmark or standard to be utilized for these KPIs. For example, there are construction industry standards for most activities, but their applicability to the location of work, its complexity, and the skill level of the workforce may not exactly fit and could lead to skewed results. For example, if the

EPC Company estimated the excavation for a large foundation based upon a sandy type of soil, but once work starts onsite and the EPC Construction Manager finds that the soil is a red clay type, then this estimate, along with the KPI, has to be adjusted for this type of soil because the excavation of clay type soil takes much longer than the removal of sandy soil. After this step is completed, the final step is to define how change management will be performed over the life of the project. The contract usually has specific language regarding when the customer must be notified once a change in the work scope has been identified and what changes qualify for reimbursement. This change management approach should include both financial and schedule impact information in the change order that is submitted to the customer for his or her review and approval, which should be done prior to the start of the work. On some projects, if there will be a significant impact on the project's budget or schedule, then the site Program or Project Manager must submit this change order to both the EPC Company and the customer's Change Control Board (CCB) for the project and wait for their decision before commencing the work related to this change order. This process is shown in Figure 2-4.

2. **Project Control** – This area of the PMP focuses on the project's milestones as they are specified in the contract and how the Program or Project Manager intends to manage the project's work schedule onsite to ensure that each milestone is being achieved at the time specified. The WBS developed for the project only includes the logical steps for the project, but it has no durations associated with each of the activities that are listed. After a duration is added to each activity, then the WBS is transformed into a project schedule with an established critical path. The Project Manager must now include what level of detail the schedule, usually Level 5, or work package level, will be updated each week onsite by each EPC subcontractor for a three-week look ahead so that the completion of the critical path activities will remain on track. This level of detail is usually not the same as what the EPC Contractor provides to the customer on a monthly basis because it is usually a Level 3 look-ahead schedule, based upon the contract, and is primarily designed to provide an overview of the project's overall progress with respect to established milestones. The other item that should be included in this section is exactly what project control meetings will be conducted onsite so that feedback from both the EPC Company's subcontractors and the customer can be received for continuous updating of the two different types of schedules. For example, an EPC Company typically establishes a weekly meeting for updating

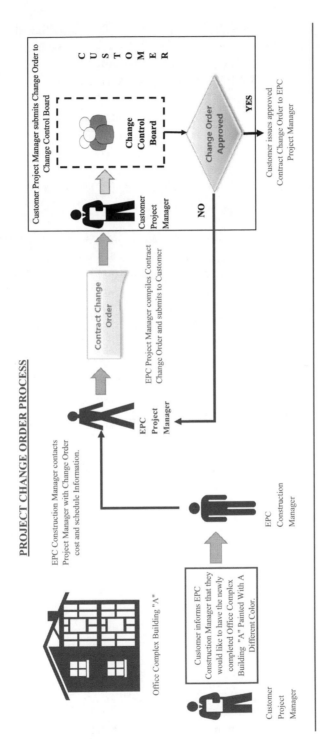

Figure 2-4 Project Change Order Process

the three-week look-ahead schedule, which only involves a scheduling person from each subcontractor and very brief daily morning meetings with just the Construction Manager from each subcontractor. These daily morning meetings are primarily held for project work coordination and support, which leads to a stronger project team and more effective management of the work being performed onsite. If the project is on track and the milestone dates are being met as scheduled, then the project control meetings with the customer are usually conducted on a monthly basis. The agenda for these meetings is usually fixed in advance, along with their location—onsite or offsite at some other location—and are designed for the discussion of any site-related issues that require site management support for resolution.

3. **Project Alignment** – In a complex construction project, the EPC Program or Project Manager will be dealing not only with many subcontractors but also with many internal department managers, such as Engineering, Human Resources, Accounting, Finance, etc. It is easy for the project to move out of alignment with the overall organizational goals of the company, which is hard to fix once it occurs, and senior management starts to "step in" per directions from the company's board members. Figure 2-5 demonstrates how this alignment should be viewed by the EPC Program or Project Manager, which must be implemented prior to the start of work onsite and constantly updated over the life of the project.

The external governance starts at the Execution Leadership Level (see Figure 2-5) at the beginning of the project, with the customer providing Engineering design specifications for all of the equipment, piping, and materials that the EPC Company must adhere to in the Engineering, Procurement, Manufacturing, and Installation processes. The EPC Company translates this requirement into many Approved for Construction (AFC) drawings, Field Quality Procedures (FQPs) for installation, and Inspection Test Procedures (ITPs) for manufacturing, which the customer approves (discussed in further detail in Chapter 6), for their contractors, vendors, and suppliers at the Delivery Management Level, as depicted in Figure 2-5. The internal communication alignment for the EPC's governance of a large, complex project should start first with the input from the EPC Company's senior management, once the Program and Project Managers have submitted the Construction Execution Plan (CEP), which is performed between the Executive Sponsorship Level and Execution Leadership Level. The next step is to ensure that each work activity in the CEP has the same WBS Number specified for it in the

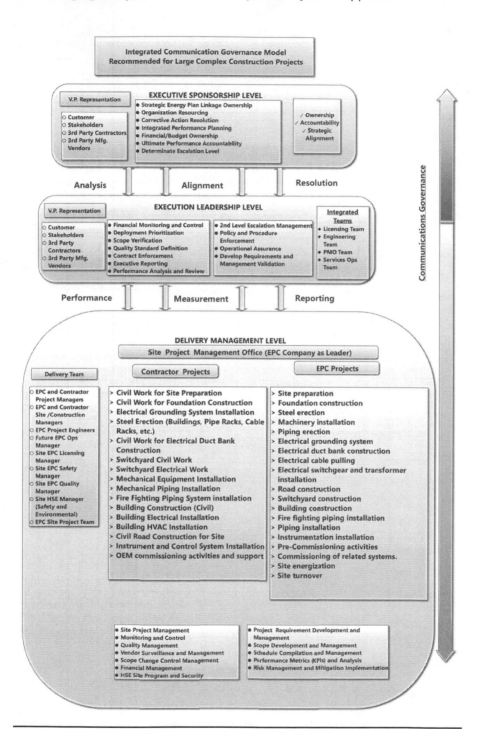

Figure 2-5 Integrated Communication Governance Model Recommended for Large, Complex Construction Projects

project schedule. After this is completed, then the appropriate WBS number for each activity should be incorporated in the document, schedule or report that is shared by all of the project stakeholders for these activities—currently not a standard practice on most large, complex construction projects. This alignment is vital to the effective communication among all three levels of the program or project, as shown in Figure 2-5.

   After this internal alignment and governance for the program or project is established, then the next step is to establish what the site will require in the area of business systems to ensure that this communication remains effective for all stakeholders and provides the right information that the EPC Company's management requires to ensure alignment is being maintained over the life of the project. For example, if one of the corporate goals is that all projects will provide a 4% margin of profit, then the Program or Project Manager has to not only establish a financial report that provides updates on a specified time interval but also immediately report any financial problems for action by management, if necessary. In addition, the Program or Project Manager must also ensure that all subcontractors are communicating the same type of financial information to them so that he or she can ensure that they will not have cost overruns that will impact the project's margin of profit.

4. **Project Budget Management** – The project usually has three basic items that must be managed by the Program or Project Manager: time, money, and quality. This area of the PMS specifies what financial systems the Program or Project Manager will implement at the time of project initiation and update at the end of each phase of the project to ensure that the project's budget is maintained. This becomes a greater challenge when the project is a complex construction type, with a large number of subcontractors and a continuous stream of consumables, equipment, tools, materials, etc., required for the large amount of work to be performed. The one area that requires particular attention is the contract with each subcontractor and what that subcontractor is responsible for when performing its work. For example, the painting subcontractor may be responsible for painting the exterior of a building, but the expense for the scaffolding erection to perform this work may come under the EPC Company's responsibility. This can be a large expense if the building is a large three-story apartment building, and it is usually not something that is added into the project's budget.

5. **Project Quality Management** – The quality of the deliverables for a complex, construction management project does not involve just the

final product but all of the materials and equipment that are installed to produce them as well. The Program or Project Manager has to establish within the PMP what system will be established onsite so that the quality standards specified in the contract will be met, which include any and all supporting documentation. The next step is to develop the site quality organization that will be responsible for performing the required quality control and auditing for compliance by all of the subcontractors performing the work. The Quality System, which will be discussed in the next chapter of this book, is critical to the PMS because its outputs are what will be required by the customer before the deliverables are accepted and the final payment to the Program or Project Manager is provided for the project.

C. **Project Quality Plan**

The rudimentary Project Quality Plan (PQP) utilized during the project presentation to the company's management for project approval now needs to be developed further to ensure that the required quality for each contract deliverable is maintained over the life of the project. The one important item to remember is that quality cannot be inspected into a deliverable but must start at the time of Engineering design. Many EPC Companies try to use a "canned" or generic PQP and make it fit all projects, which is the reason why many complex construction projects encounter numerous quality issues. For example, the design of a foundation for a building must comply with specific Civil Engineering standards, which include not only the excavated, backfilled, and compacted area but also the quality of the concrete, the rebar for the foundation, the erection of the ironwork prior to the pouring of concrete, the grade of concrete used, etc. A mistake in any one of these areas can lead to the foundation being rejected, its removal, and starting over, which can have a large financial and schedule impact on the project because it is all rework at the EPC Company's expense. The PQP must establish a Quality Control and Assurance System (discussed in Chapter 6) for the project that will guarantee that situations such as this will not happen. The steps to building a solid PQP are as follows:

1. Define exactly the purpose of the PMP and identify each of the project's deliverables along with how their quality will be maintained over the life of the project.

2. Establish the Quality Organization that will be onsite where the construction will be performed, along with the roles and responsibilities of each position.

3. Compile an audit schedule for the life of the project for each of the phases of the project, which for an EPC Company are the following:

a. **Engineering Design** – This phase involves the development of the drawings that will be used for the construction and production of the deliverables.

b. **Manufacturing** – This phase involves the manufacturing of the required equipment, structural steel, various construction materials, piping, etc.

c. **Construction** – This phase involves the actual erection of the buildings and structures, followed by the installation of the specified building services.

d. **Commissioning** – This phase involves the testing and operation of the various systems, as required, to produce the required deliverables, such as heating ventilation and air conditioning (HVAC), lighting, water, solar water heating, chilled water system, etc.

e. **Turn Over** – This phase involves turning over the finalized deliverables to the customer, which includes all required documentation.

f. **Close Out** – This phase involves closing out contracts with all of the various subcontractors; any contract change orders, which the customer has signed during the construction process; and receiving the final payment from the customer.

4. Identify all stakeholders and what their communication requirements are regarding quality so that the Quality Control and Assurance System developed can provide the right information at the right time. This should include meetings, notification of deficiencies, reporting requirements, etc.

After this Quality Plan is completed, the Project or Program Manager should provide a copy to the customer for review and comment since the customer will also have a similar Quality Organization onsite working for them.

D. **Project Financial Register**

In a complex construction project, the Program or Project Manager usually encounters situations that call for a change order to the contract, which has to be approved by the customer. At the same time, however, they can be given a change order by their subcontractor. A change order from one of the EPC's subcontractor has high priority because, in some cases, it can lead to a stoppage of work until it is negotiated and approved by the EPC Program or Project Manager. The Project Financial Register is critical because it not only provides a way of tracking these various change orders but also the following important items for the project:

1. A weekly or monthly report of all change orders along with their financial impact on the project's budget, which senior management needs to know, along with the company's accounting department.

2. A financial checklist for the Program or Project Manager to use at the time of project close out so that all financial matters are closed for the EPC Company, the customer, and subcontractors.

An example of a Project Financial Register is shown in Table 2-5.

Table 2-5 An Example of a Simplified Project Financial Register

Project Name:	Project Location:	Project No.:	Project Start Date:	Project End Date	Customer Name:
Alpha Center	Peoria	ACME-0021	09-12-15	01-12-17	XXXX
Commercial – Contract Change Orders					
WBS No.	WBS Description	Change Order No.	Change Order Description	Change Order Cost	Change Order Status
2.2.2	Paint of exterior wall	001	Changed Color	$X,XXX	Submitted
Internal – Subcontractor Change Orders					
WBS No.	WBS Description	Subcontractor Name	Change Order Description	Change Order Cost	Change Order Status
3.2.4	Install roof truss	O-Roofs	Drill holes for mounting	$XXX	Approved

The WBS number for the associated schedule activity is critical because it assists the Program or Project Manager in determining if this change order affects the critical path of the project, which means that the change order must provide both a cost and a schedule impact for proper evaluation. If the Project Financial Register is set up in a spreadsheet format, then the Program or Project Manager can easily set up a search function so that the information required can be readily found when needed.

E. **Project HSE Plan (Health, Safety, and Environmental)**

The Health, Safety, and Environmental Plan (HSE Plan) for the project outlines how the health and safety of all onsite personnel will be ensured and maintained throughout the construction phase of the project. The main objective is to not only guarantee that there will be no lost time from accidents but also that all reporting and safety guidelines as specified in the contract will be followed. This extends to the environmental aspects of the project such as effluent discharge from the site into the local waterways, disposal of any and all chemical wastes, storage of chemicals used onsite, etc. In order to accomplish these goals, the Project HSE Plan is usually broken down into the following areas or sections:

1. **Statement of the Project HSE Plan's Objectives and Contractual Obligations** – This statement is necessary to ensure the safety of all personnel onsite and to comply with all local HSE ordinances, which are only two of the objectives specified in the HSE Plan. This area should cover topics such as substance abuse, right to stop work, site smoking policy, driving onsite speed limitations, specific safety rules that must be followed, etc. The legal requirements of not only the EPC Company, but also the local government, should also be clearly stated in this area of the HSE Plan.

2. **HSE Organizational Chart** – This organizational chart for the EPC Company, all associated subcontractors, and the HSE Department will be established onsite to accomplish all of the items mentioned in the first section of the HSE Plan. It should also provide the responsibilities of each role shown in the organizational chart.

3. **Permit-to-Work (PTW)** – If a subcontractor for the EPC Company wants to perform work, it must have a PTW from the EPC Company that provides the following information:

   a. The geographical area on the construction site where the work will be performed

   b. The applicable names of the persons responsible for supervising the work in this area, along with the hours that will be worked

   c. The nature of the work and the safety hazards are inherent in this type of work.

   d. The safety methods that will be employed to control the identified safety hazards and all emergency numbers to be used in the event of an accident

   e. The signature of the subcontractor's Project Manager and responsible supervisor confirming that the information provided is accurate and that the safety program requirements of the EPC Company will be followed.

   If a subcontractor is found working onsite without the required PTW, usually its work is immediately stopped, and disciplinary action is taken against the subcontractor's Project Manager to ensure that this will be the only violation. The subcontractor's performance is monitored on a weekly basis to ensure its compliance.

4. **Hazardous Chemical Plan** – This area will lay out what steps will be followed to ensure that all Hazardous Chemicals, such as cleaning chemicals, paints, solvents, etc., will be safely stored, disposed of once used, and handled onsite. The one standard requirement is that all Hazardous Chemicals brought to the site must have a Safety Data Sheet, which is provided by the manufacturer or distributor of the

particular chemical, before it can be brought onsite. The Safety Data Sheet for the user of the chemical provides the following information:

a. The name, address, and phone number of the manufacturer or distributor and the regular uses of the product

b. The identification of the hazardous waste along with its particular class, such as a flammable liquid, corrosive liquid, epoxy resin, etc., as well as, in words and pictorially, a warning describing the hazards associated with this product

c. The chemical breakdown of the product along with common names for it in the industry

d. The first aid measures that should be followed, depending upon the method of exposure to the chemical and what symptoms can be expected when exposure occurs

e. Any firefighting measures that must be taken for the proper handling of the chemical, which includes special protective personal equipment and the safest methods to employ in the event of chemical burns

f. The correct methods that must be employed for proper storage and handling of this particular chemical, which also includes its disposal and any measures to be taken in the event of an accidental release

5. **Daily Toolbox Talks** – This section will describe what type of daily toolbox talk should be held with the workers, which will be conducted at the beginning of the shift or at the start of work activities. This form of communication is mandatory so that the safety hazards of the work area are known by all personnel. It will also include a form that each subcontractor must fill out and sign prior to its submission to the EPC Company safety officer.

6. **Safety Training** – This section will list what type of training, on a weekly basis, will be conducted for the subcontractor's frontline supervisors, in addition to the other specialized training that will be provided, such as working at heights, confined space, rigging, etc.— usually as the situation arises.

7. **Lifting Equipment, Rigging, and Lifting Procedures** – This section details not only the lifting procedures that will be employed onsite but also the inspection of the rigging involved with the lifting and the lifting equipment. This lifting equipment can be mobile cranes, crawler cranes, boom lifts (personnel type), tower cranes, etc.

8. **First Aid and Medical Treatment** – This section will specify what local medical facilities will be utilized in the event emergency medical aid is required and what immediate first-aid facilities will be established

onsite for non–life-threatening injuries that may occur during the construction onsite.

9. **Hand Tool and Site Equipment** – This section will explain how the electrical hand tools, power cords, and all operating equipment onsite, such as hydraulic pipe bending machines, rebar benders, cutting gas rigs, etc., will be inspected on a monthly basis. It should contain a color-coding system so that it can be readily confirmed that each tool or piece of equipment has been inspected and is safe for use.

10. **Communication Plan** – This section establishes how the HSE communication will be established onsite between the customer, the subcontractors, and the EPC Company. It should include details of a Safety Kickoff Meeting at the beginning of the project for all parties involved and establish an onsite HSE Steering Committee, which will comprise representatives from each of the subcontractors, the customer, as well as the EPC Company. These members will also be involved in a weekly site walk down so that any safety hazards are readily resolved at that time. In this section, notification protocols for safety incidents along with reporting requirements should also be clearly explained, along with the necessary forms that will be utilized for this notification.

11. **HSE Inspection and Audit Plan** – This section will detail the HSE Audits that will be conducted on each subcontractor prior to its start of work onsite as well as what daily, weekly, and monthly inspections will be performed onsite to ensure that all personnel and work areas are in compliance with all site-specific HSE rules.

12. **Site Signage and Traffic Control** – This section will specify what type of HSE signs will be posted onsite and their exact locations. It should also establish the site speed zones and how they will be enforced by the EPC Company's HSE Department.

2.4 Concluding Remarks

The Project Management System (PMS) is complex, but when a Project, Program, or Construction Manager is managing a large, complex construction project for an EPC Company, it is necessary to effectively manage and control the large amount of work and subcontractors required for delivery of the project on time and under budget. In the next chapter, the establishment of the PMS will continue, with a detailed discussion of Part II of this PMS, which will move our discussion into the construction-execution process.

Chapter 3

Project Management System: Part II

3.1 Project Management Outputs

3.1.1 Project Construction Execution Plan

The Construction Execution Plan (CEP) is usually compiled by the site Construction Manager assigned to the project and should be developed during the initiation and planning phases of the project. It is designed to describe not only each major work activity but also how it will be performed and controlled to ensure that its quality meets the standard specified in the contract. The other important item is for the Program or Project Manager to establish alignment between Engineering, Procurement, and Construction, which can be done with the CEP during its development. The organization of a project CEP, which is for the construction of a large power plant, is shown in Figure 3-1.

The simplified organization shown in Figure 3-1 is actually only a small part of a well-written CEP because many large construction projects are complex. All of the areas shown in Figure 3-1 typically require a very large number of subsequent activities, which is why many companies usually use Subject Matter Experts (SMEs) to assist the Construction Manager in the organization of the CEP. The alignment for a CEP can be achieved by applying the respective WBS number for each item from the project's schedule baseline WBS to those shown in Figure 3-1 (discussed in Chapter 2, in the Project Alignment section of the Project Management Plan). For example, if the WBS for the Civil

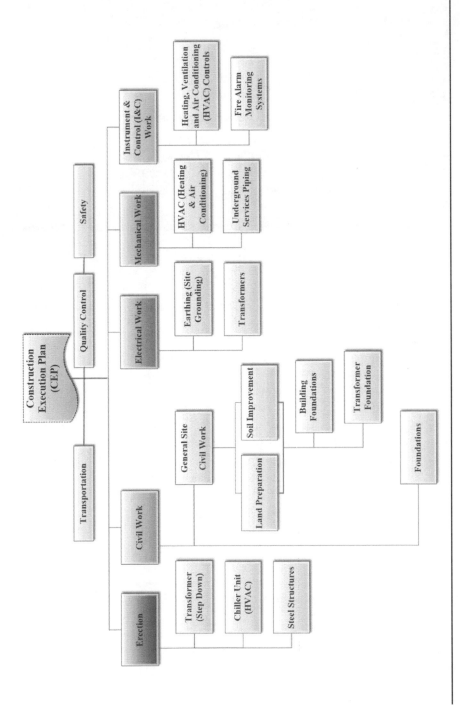

Figure 3-1 Construction Execution Plan (CEP) Organizational Chart

Work block is 1.0, then the Foundation block shown in the organizational chart would have the number 1.2 because it does not start until the Civil Site Work is completed. This alignment is critical to the effective management of the construction work by the EPC Construction Manager and should be established at the time the CEP is developed. The steps required to develop the PROJECT CONSTRUCTION EXECUTION PLAN (CEP), as shown in Figure 3-2, will now be discussed.

Flowchart

Step 1 – Development of an Overview of the Construction Area

In order to get started, the site Construction Manager needs to obtain a large General Arrangement Drawing for the construction site from the Engineering Department and then look at the area around the proposed Construction Site. The purpose for this is to not only visualize the project but also to look at the surrounding infrastructure that will be required to support the project. The items to look for are the following:

- **Site Access:** In a large, complex project, it is always a best practice to have all trucks and construction equipment come through one gate and passenger vehicles such as SUVs, vans, etc., through another gate. If the General Arrangement Drawing does not show this, then evaluate it to see if access can be improved in this manner with Civil Engineering.
- **Concrete:** Evaluate how much concrete will be required for this project, which the Civil Engineering team can provide, and then find out what concrete suppliers are readily available in the area. After their location is confirmed, evaluate how far they are from the site. This distance is critical because a major concrete pour mandates that the concrete can only be held in the truck for a short period of time before its fluidity is compromised. The next step is then to contact each of the chosen concrete suppliers and find out if they can produce the grade of concrete required by Civil Engineering for all of the major foundations, which is a critical item to know before the project's work starts.
- **Local Roads:** The first thing to find out is where the equipment and materials will be delivered. If it is straight to the construction site, and the roads are adequate for transporting all of the equipment, then there should not be any problems. However, if it is to a local seaport, then an effort must be made to determine if this seaport can handle what will be shipped, and if the local roads, bridges, transmission lines, etc., can handle the heaviest load. If the local roads or areas will need to be upgraded,

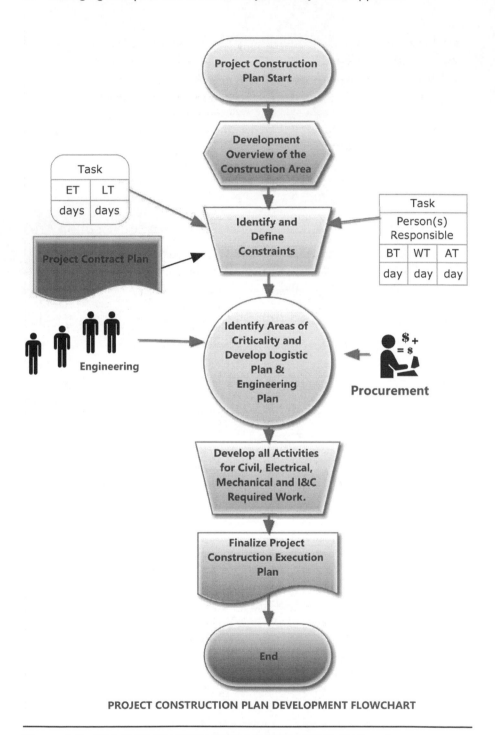

PROJECT CONSTRUCTION PLAN DEVELOPMENT FLOWCHART

Figure 3-2   PROJECT CONSTRUCTION EXECUTION PLAN Development

then the customer should be notified to take immediate action, which may require a different route, so that the equipment and materials will arrive onsite as scheduled. The tendency in most companies is to find a local heavy haul subcontractor and then leave everything to the subcontractor, which does work in most cases because they are familiar with the area and local ordinances, but the Construction Manager should always follow up on this item with the selected subcontractor.

- **Road Restrictions:** The Construction Manager should also review the project's delivery schedule in relation to the local area's holidays because there may be load restrictions in place at that time by the local authorities.
- **Site Effluent Drainage:** The one area usually overlooked on most projects is the construction site drainage, which may require holding ponds for sediment settling before it can drain into the local waterways. These local waterways have specific local restrictions for discharge of effluent waste, and failure to comply can lead to revocation of the construction site work permit. The Construction Manager needs to establish a site drainage plan at this point, so that these systems are in place prior to the start of work.
- **Laydown Area:** It is critical that a laydown area for the equipment and material, which will be brought to site before their installation, is evaluated and established prior to the start of work. This area has to be large enough for the loading and unloading of these items, which on many projects becomes a large logistical nightmare if the area needs to be at a very far distance from the construction site. These evaluations must also include the type of security needed—such as fences, security guards, etc.—that will be required before any of the equipment and materials arrive.
- **Site Support Services:** The Construction Manager must review the Site General Arrangement drawing and determine where the best place will be to establish site offices along with potable water, electricity, and toilet services for the EPC construction staff and the customer's site personnel. The one area to determine is if internet will be available, and if not, then what type of temporary solution can be provided until such service is available. There will be thousands of documents and drawing transmittals, and internet service is required so that the correct drawings are onsite for the start of construction.
- **Site Waste Materials:** In a large, complex construction project, there is usually a large volume of existing soil that must be removed in order to start the process of foundation erection for buildings, machinery, structures, etc. The Construction Manager needs to contact the customer and find out where a place is located that can accept all these construction spoils. At the same time, knowing the distance and time, along with what permits are required, is also critical for planning purposes.

Step 2 – Identify and Define Constraints

This step requires the Construction Manager to meet with the planning/ schedule department and the contract department to perform the following items to establish what constraints the construction project may face once work starts.

- **Activity Duration:** This item requires the planning department to perform a forward pass of the project's Level 4 or 5 CEP baseline schedule to determine the Early Start and Early Finish dates for each activity. The next step is to perform a backward pass to determine the Late Start and Late Finish dates for each activity. This will establish the most optimistic duration and the most pessimistic duration for each activity. After this is done, an analysis can be performed to see if the "as planned" durations within the baseline schedule either fit or do not fit within this window. For example, if the critical path activity, "Start excavation for Building A," in Table 3.1 shows an Optimistic Duration of 15 days and a Pessimistic Duration of 26 days, then the Planned Duration of 22 days must be changed because the large difference between the Optimistic and Pessimistic Durations for this activity suggest that there needs to be more than 4 days of float for this activity in the schedule.

Table 3-1 An Analysis of the Duration for "Start Excavation for Building A"

Activity No. and Description	Optimistic Duration	Planned Duration	Pessimistic Duration	Comments
1.1.1.12 – Start Excavation for Bldg. A	15	22	26	The planned duration needs to be less because it only has four days of constraint.

The reason for the comments in Table 3-1 is that if this is a critical path activity, it can only slip four days before it starts to push out the schedule, which has a high risk of occurrence because there are many unknowns associated with excavations, such as hard soil, underground water, etc. The Construction Manager should develop a list of all activities on the critical path of the baseline schedule that are constrained in this way, and then move to include those activities that are sitting 30 to 60 days off of the critical path on this list. This provides the Construction Manager and the Program or Project Manager with a list of the areas where the schedule is constrained and has a high risk of pushing out the schedule, which they should work to improve with more equipment, labor, etc., prior to the start of work. For example, in the case mentioned above, increasing the number of excavators and trucks for this activity will reduce its duration and remove this four-day constraint, which will reduce its risk to the project's baseline schedule.

- **Contract Constraints:** This item involves working with the Contract management group to evaluate the Project's Contract Plan and determine what contractual constraints, internal or external, may impact the construction work. Some of these constraints can be:
  o **Site Permitting** – This may require assistance from the customer because it may involve filing and receiving approval from both the local municipality and the government for the construction site permit, which can be time consuming and delay the start of work.
  o **Local Workers** – There may a clause in the contract that requires specific skills, or as another example, a specific percentage of the workers must be from the country where the project is located, which impedes the staffing process for the project and can cause delays to the project's schedule, if these particular skills are not available at the time they are required onsite. If this clause extends to engineers, then the question of salary and what their responsibilities will be could be another constraint for the project. This constraint usually manifests itself in additional costs that were not included in the project's initial budget, and it reduces the skilled staff that the EPC Company planned to bring onto this project.
  o **Environmental** – This contract may have a clause that specifies strict compliance regarding any environmental impact on the surrounding waterways, ocean, and any areas affected by the project. This compliance may lead to constraints if the required environmental control systems are not in place prior to the start of work and may require the additional cost of a contractor to be brought onsite for this purpose. For example, the effluent runoff from the construction site during a rain shower can include large amounts of various sediments, which must be filtered prior to this water being released into the local drains.
  o **Progress Payments** – In most contracts, milestone payments are usually established. The Construction Manager, along with the Program or Project Manager, must analyze these payments in relation to their planned cost schedule for the project. If the cost to achieve the earlier milestone payments is high, which is the standard case for large, complex projects because civil work requires large amounts of machinery and material, then the associated milestone payments must cover these

costs. If they do not, then the project will be financially constrained in the beginning of the project, leading to project delay because the mobilization of additional resources or equipment may have to be postponed for funding purposes, which would lead to a budget overrun and project failure.

○ **Payments** – Free-On-Board (FOB) is a term used primarily in the international arena to specify exactly where the customer takes ownership of the equipment or materials. If it is at the port of entry into the country in which the construction project is located, then the Construction Manager needs to understand two things, as follows:

➢ Where the port is located and how far it is from the construction site, which will affect the transportation of equipment and material to the project

➢ If the customer has to provide payment at the port prior to customs clearance and release for unloading, which can delay the release of equipment for lack of payment

These two conditions can cause a major delay and put a constraint on the flow of equipment to the construction project when it is needed for installation.

After these constraints are known, the Program or Project Manager, the Construction Manager, and the Contracts Manager should establish a program to manage each known constraint. The Project Contract Plan should be updated with this Procurement Plan, and it should be incorporated into the CEP, which will link all three documents and ensure the work being performed onsite is in accordance with the contract.

• **Critical Areas:** The first item the Construction Manager needs to get is a complete Level 5 or the Work Package level logic network map from the Planning Department. This Level 5 schedule logic network map will not only show the flow of work activities but also the interrelationships between the civil, mechanical, electrical, and I&C discipline work that will be performed onsite. After this is completed, the next step is to have a number of meetings with the following departments to develop a Logistical and Engineering Plan that matches the sequence of work onsite:

○ Engineering
○ Logistics
○ Procurement

It is critical to have the department managers in these meetings, along with all of the engineers, logistics personnel, and procurement personnel that will be assigned to the project. The focus of these meetings is to identify the following critical areas so that all parties are working toward the same schedule:

○ **Approved for Construction (AFC) Drawings** – The establishment of when Civil, Mechanical, Electrical, and Instrument & Control (I&C) Engineers need to approve drawings to start construction is essential to starting and finishing the work as scheduled. These dates must also include enough time for the customer's engineers to complete their approval once these drawings are submitted to them from the EPC's Engineering Department. The Program or Project Manager must ensure that the schedule developed in this area is maintained over the life of the project. The members of the procurement group must also know this schedule so that they can make sure that the purchase orders to various vendors are issued with enough time to support the scheduled delivery at the construction site. Because the logistics group will be involved with shipping and customs clearance, they must also know the factory "release for shipping" dates for each piece of equipment and material.

○ **Project Engineering Plan** – After the critical dates are established, along with all related interface issues between the various Engineering disciplines, the Engineering Department must submit its Project Engineering Plan. The Construction Manager should then go through this plan to verify that the dates and the proposed drawing issuance priorities for each discipline matches how the work will be performed onsite. If there are changes to be made, now is the time to make them and not when the work has already started onsite. This will also help the Construction Manager to minimize rework or any delays that are due to lack of the "AFC" drawings onsite. The Program or Project Manager for the EPC Company needs to ensure that this Engineering Plan is followed and updated immediately if the customer wants something changed. This revision process must be closely monitored so that it remains in alignment with the established baseline schedule and the CEM established by the Construction Manager.

○ **Project Procurement Plan** – The Procurement Department manager must now have the team associated with the project start contacting the various equipment and material providers to verify that they can meet the specified critical dates. A commitment from each of the providers must be in writing, and any changes immediately relayed to the Construction Manager for action.

○ **Project Logistics Plan** – After all changes have been made to the agreement by the Construction Manager, the Engineering Department and Procurement Department, the Logistics Department manager will prepare a logistics plan for the project. This plan must include the places where the items will be shipped and the port for receipt of all items, along with any and all expected customs requirements for clearance at time of receipt. If the customer will be required to expedite this customs clearance, then the Program or Project Manager for the EPC Company will be tasked with contacting the customer with these requirements and monitor this commitment once the project work starts.

o **Revisions** – All parties under the direction of the Program or Project Manager must establish a communication plan for any and all revisions that may be made to any of the three plans and the CEM. This revision communication plan must be concise, accurate, and support the project's baseline CPM schedule. It must also be incorporated into the Project Communication Plan (PCM) and include approval from the customer representatives so that they are aware of their requirements.

o **Risk** – The last critical area is to compile from the engineering procurement and logistics departments a list of qualified and quantified risks that may affect the CEM and the baseline schedule. The Program, Project, and Construction Managers should take this information and develop the necessary mitigation strategies such that all departments feel that they can accept. This completed CEM Risk Management Plan should not only be in the Construction Execution Plan but also added to the Project's Risk Management Plan, which was discussed earlier. Once the work onsite commences and over the life of the project, both of these risk management plans need to be continually updated.

Step 3 – Compile and Develop all Activities for Civil, Electrical, Mechanical, and I&C Work

This step will require the Construction Manager, with assistance from various sources, to now build the CEP for all of the construction activities that will be required for the production of the project's deliverables. In the CEP, each area under the general heading will have a detailed explanation of what work will be performed in that area, how that work will be done, what specifications will be used, and the resources required for this work. The intent is to describe not only each major work activity but also how it will be performed and controlled so that the quality of work meets the standards specified in the contract. The organization of a CEP for a large, complex construction project is graphically depicted in Figure 3-1, which appears at the beginning of this chapter.

Compilation Process

For example, the area, "Land Preparation," depicted in Figure 3-1, would be written and constructed in the following manner for a building foundation:

Land preparation work for the building foundation will be performed by first reviewing the site survey and then excavating each area to the level specified in the Approved for Construction (AFC) drawing for this particular area. This work will involve at least three to four large excavating machines and a large number of dump trucks, as required, to meet the durations established in the project's schedule for this activity. The personnel required will primarily be a small crew of twenty laborers and operators for the machines.

The next step for the Construction or Project Manager is to translate this statement into the actual duration and cost for each item, which is shown in Figure 3-3.

The one cost, not reflected in the numbers shown in Figure 3-3, would be the cost of a site survey team, which is normally provided by a contractor and used for all of the civil activities performed onsite. This site survey team is required for all excavations to monitor and verify when the final excavated level has been achieved per specified civil drawings. This approach minimizes the impact of cost on the project, but the Project Manager has to ensure that the contractor can provide enough personnel to cover all of those civil activities being performed at the same time. The next step in the CEP is to take this "Land Preparation" format for a building foundation and apply it to the other specific areas or foundations that will be need to be developed, which will provide an exact number of each machine and the resources required to effectively complete all of the land preparation work onsite.

This process sounds easy, but the development of an effective CEP requires the person or persons to take a final product (i.e., a foundation) and mentally work backward to identify all of the required steps in the correct logical sequence. The CEP, after it is compiled in this manner, now becomes the work execution plan for the Project and Construction Managers to follow over the life of the project. In addition to a work execution plan, it is also an accurate source of information for the compilation of the project's budget and schedule, which must be reviewed periodically by the Project Manager because conditions on the construction site can change.

The other advantage of a well-constructed and effective CEP is that it can assist the Construction Manager in overall site access planning so that the logical sequence of the work will not impede the flow of equipment and critical resources to areas when they are required. If just one foundation, as shown in Figure 2-8, can require a large amount of truck traffic for over six months, then imagine the impact of two similar foundations or more being constructed in the same area at the same time.

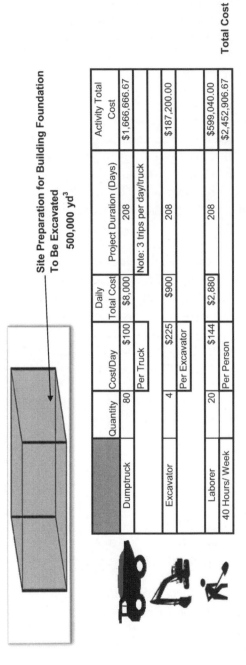

Site Preparation for Building Foundation To Be Excavated 500,000 yd^3

	Quantity	Cost/Day	Daily Total Cost	Project Duration (Days)	Activity Total Cost
Dumptruck	80	$100 Per Truck	$8,000	208 Note: 3 trips per day/truck	$1,666,666.67
Excavator	4	$225 Per Excavator	$900	208	$187,200.00
Laborer	20	$144 Per Person	$2,880	208	$599,040.00
40 Hours/ Week					$2,452,906.67

Total Cost

Figure 3-3 Cost and Duration Breakdown for Site Preparation Activity

Step 4 – Alignment

The current problem with most CEPs is that they are primarily focused on cost, which usually precludes the activities being linked to the completion of specific milestones. However, if at the time of project planning, the project's Work Breakdown Structure (WBS) is linked with the project's CEP, as shown in Table 3-2, then the Project Manager will have a powerful tool for effective management of the construction project because its work activities will be linked and aligned to specific milestones for the project deliverables.

Table 3-2 WBS Assignment of Activities Shown in Figure 2-7

Construction Activity	WBS No. Assigned	Engineering Discipline
Civil Work	1.0	Civil
General Site Civil Work	1.1	Civil
Land Preparation	1.1.1	Civil
Building Foundations	1.1.1.1	Civil

The Construction Manager will also be able to evaluate the impact on the schedule quickly in the event that an incident occurs, such as equipment breakdown, worker strike, heavy rains, etc., during the execution of one of these activities. The other advantage is that if one activity is starting to get close to its projected cost before it should be done, then the Construction Manager can immediately contact the contractor's Construction Manager for action.

Step 5 – Finalize Project Construction Execution Plan

In this final step, the Construction Manager will finish the CEP and then submit it to the Program or Project Manager for approval, which also may include the management of the EPC Company. The purpose for this company management approval is to establish ownership at all levels within the company for support and clarity regarding progress reports that the Construction Manager via the Program or Project Manager will be submitting on a specified timely basis. It should be noted that, in some contracts, the customer may stipulate that work cannot start work onsite until the EPC Company has an approved the CEP for the project.

3.1.2 Project Engineering Plan

The Project Engineering Plan (PEP) is of paramount importance to both the Planning and Execution Phases of the project. In the execution phase, construction work onsite cannot start without the Approved for Construction

(AFC) or Issued for Construction (IFC) drawings being onsite. In the planning phase, procurement cannot start until an engineering drawing is approved, and the corresponding Bill of Material (BOM) from that drawing is given to the Procurement Department. It is the responsibility of the EPC Company's Engineering Manager to ensure that the PEP is developed once the contract is awarded and updated in accordance with the CEP's baseline CPM schedule for the delivery of each engineering discipline's drawings, which usually number in the thousands. The PEP should be established in the following manner:

- Define the purpose of the PEP and how it will support the project, which should include how its goals will be kept in alignment with all required corporate goals.
- Display an organizational chart for the Engineering Department that will support the complex construction project, which should include the name, email, and phone number for each person. On large, complex construction projects, this chart should include a site Engineering Manager and a small staff of design engineers to support any design-change requests onsite. The responsibility of this group onsite will also be to issue the Field Design Requests (FDR) to the customer and monitor their status after submittal so that there is no impact on the project.
- Define the roles and responsibilities of each of the persons shown in the organizational chart and ensure that their area of responsibility is clearly defined. For example, do not say the mechanical engineer is responsible for all Mechanical Engineering issues onsite, but rather state this person is responsible for all Mechanical Engineering issues as they relate to piping and respective piping supports onsite. This approach makes it easier for the Program, Project, or Construction Manager to quickly locate the correct engineer when a piping design issue occurs onsite.
- Compile a Level 5 CPM (Critical Path Method) or work package level schedule for drawing submission to the customer that matches the requirements of the CEP and the project's CPM baseline schedule. This same schedule, approved by the EPC Company's management, the Program or Project Manager, and the Construction Manager, should also be submitted to the customer so their Engineering group can prepare their Engineering staff accordingly.
- Develop a **Master Drawing List (MDL)** that will clearly specify how the EPC Company will incorporate all approved drawings into an MDL database, which must include how access to it by all stakeholders will be controlled by the onsite Document Control Center.
- Develop a communication plan that graphically shows what process is to be followed for any Field Design Requests (FDR) submittals from

the construction site and how all stakeholders will be updated regarding Engineering drawing issuance, including any revisions. This plan should also ensure that it incorporates the communication requirement specified by the contract regarding submission of FDRs, once approved, to the customer's Engineering group for their subsequent approval. The other area to address within this plan is how Engineering will submit Material Approval Requests (MARs) to the customer for his or her approval, and the timing requirements for approval once the customer receives the MAR.

* Develop and establish a Project Engineering Quality Plan (PEQP) so that once an approved drawing being used onsite has been revised and issued, the related outdated drawings will be removed immediately and replaced with the newly revised drawing, after it is issued. This quality plan must clearly show what process will be followed onsite for this removal as well as the audit schedule for the MDL, so that it shows the latest revision of each approved drawing on a daily basis. This PEQP must also list all of the specified codes and standards that will be followed during the design phase for each of the Engineering disciplines, which should also be in accordance with those specified in the contract by the customer.

After the PEP is completed by the Engineering Manager and his or her staff, it should first be submitted to the Construction Manager and the Program or Project Manager for their review and approval. Once this approval is received, then the PEP should be submitted to the EPC Company's project sponsor for corporate approval prior to its implementation, which must be completed during the planning phase of the project and prior to the start of work onsite.

3.1.3 Project Procurement Plan (PPP)

On a large, complex construction project there is a massive amount of material, along with equipment, which must be correctly ordered, tracked, and shipped to the construction site. This is the responsibility of the EPC Company's Procurement Department. The standard process followed for large, complex construction projects is shown in Figure 3-4.

The customer must always be consulted about any vendor or material selection because these items are all part of the project deliverables that the customer must accept when the work is done. This responsibility also applies to the site Program or Project Manager when it comes to contractors bringing material onsite that will be permanently installed such as floor tiles, HVAC ducting, bathroom fixtures, etc. The other responsibilities for the Procurement Department are the following:

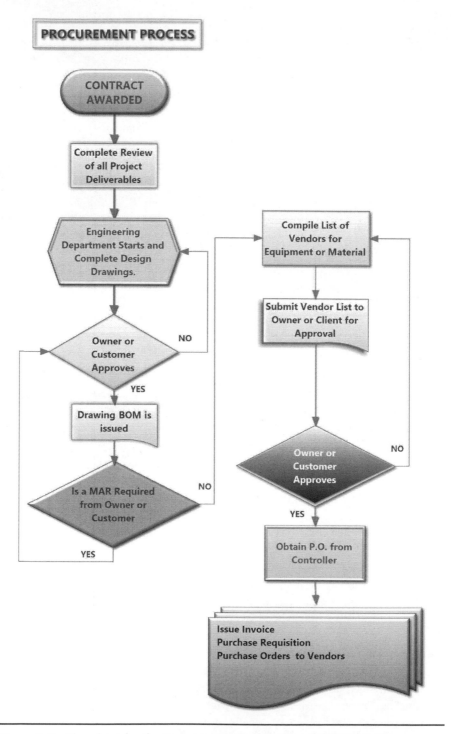

Figure 3-4   Flowchart for the Procurement Process by an EPC Company

a. **Country of Origin Verification** – The customer usually stipulates that he or she will not accept any equipment or material manufactured in certain foreign countries, which if discovered can lead to the rejection of this equipment or material by the customer. The Procurement Department must make sure that all selected vendors are aware of these requirements and are in full compliance throughout the manufacturing process.

b. **Factory Notifications** – In the manufacturing of equipment and materials, there are many points at which the customer's representative must be present to witness and sign off on the respective ITP (Inspection and Test Plan) hold points. The date, location, and time for each of these points must be communicated to the customer by the Procurement Department. Failure to perform this notification can lead to the particular material or equipment being shipped late to the construction site because the customer did not approve its release for shipment can have a major impact upon the project.

c. **Shipment Schedule** – The Procurement Department must continually monitor all vendors to ensure that their promised shipment dates do not slip and remain in accordance with the site requirements set forth in the project's CEP, which are also in line with the project's Level 5 CPM baseline schedule. The Procurement Department must also provide a monthly schedule for all equipment and material to all stakeholders, which includes the EPC Company's Logistics Department, so that everyone is aware of when the equipment or material will be ready for shipment.

After the equipment or material is delivered to the site and payment has been received, the Procurement Department must follow up to ensure that the vendor submits all required documents for payment within the time frame specified by the EPC Company's Finance Department. The Program or Project Manager for the project should ensure that they are receiving reports on this process so that there will not be any outstanding invoice problems at the time of project closure.

3.1.4 Project Logistics Plan

The Project Logistics Plan (PLP) specifies not only how each piece of equipment or material, such as structural steel girders, trusses, beams, etc., will be shipped from each location where it is manufactured but also what port will be used for entry into the country in which the construction project is located. In this plan, the Logistics Department needs to specify and perform the following:

a. The shipping companies that will be utilized and what ports will be used for shipping each piece of equipment or material from the manufacturer or supplier to the construction site.

b. The point where the customer will take receipt of the equipment or material and how the payments need to be structured for customs clearance upon its arrival.

c. A detailed logistics organizational chart for this complex construction project that clearly shows which person has responsibility for which piece of equipment or material. This will help the Program or Project Manager at the construction site to know whom to contact should any problem arise during the shipping of the item or items or with customs clearance at the port of entry. The logistics person has to understand that he or she must follow each piece of equipment or material once it is shipped and continue this support until it has arrived at the construction site, as well as ensure that all required documents required for clearance are complete, as specified by the customer and the contract.

d. A shipping schedule for all equipment and material that is in accordance with both the CEP and the Level 5 CPM schedule, which is updated as each item is manufactured or sourced. This shipping schedule should be updated on a monthly basis, and the logistics department must immediately flag any item that starts to slide so that the Program or Project Manager onsite can quickly take action. The logistics department must also be aware of the local holidays that can result in either the shutdown of the port of entry or restrict movement along the local roads to the construction site, which can impact the item's delivery date.

e. If any item is so large that a special local heavy haul contractor is required, then the logistics department must establish a contract with this contractor and coordinate this shipment with the Program or Project Manager onsite.

For a large, complex construction project, there are hundreds of items that must be shipped in addition to just the large pieces of equipment or material, and the Logistics Department has to manage all of them, which has to be done with close monitoring over a long period of time. If the EPC Company can establish a database that its suppliers can readily access to provide updates regarding the manufacturing cycle, then this difficult task can be made much easier for all parties involved. An example of such a database and the communication process is shown in Figure 3-5.

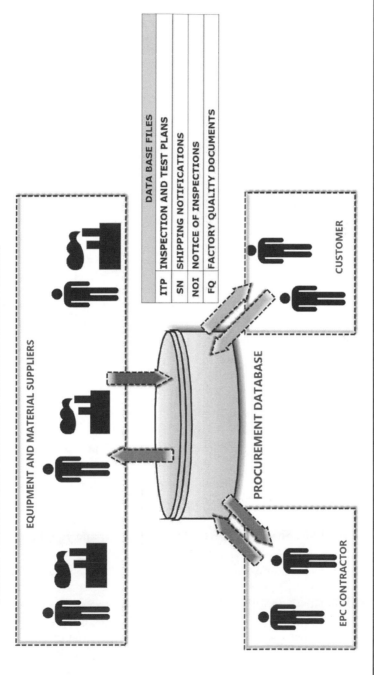

Figure 3-5 Project Procurement Database Communication Process

3.1.5 Project Communications Plan (PCP)

This Communications Management Plan establishes a communications framework for the large, complex construction project, which has much different communication requirements that are due to standard multiple-office locations. For example, the first office, the "EPC Company Office," is usually located in its parent country, which is typically far away from the Site Construction Office. The customer's main office is also generally located in a large city that is usually a fair distance from the Site Construction Office. In most remote construction sites, the current communication technology situation onsite is primarily an intermittent internet for all companies, with meetings—verbal and email—as the primary communication between all parties who work in the temporary portable offices at the construction site.

It is essential that a communications plan is established to identify and define the roles of people involved in the large, complex construction project, which will be through a project team directory that will provide contact information for all stakeholders directly involved in the project along with their role in the project. It will also include a communications matrix, which will map the communication requirements of this project. Requirements that will then lead to the communication methods and processes will be employed at the construction site to effectively satisfy each requirement and improve the effective management of this project onsite.

Communications Management

The **Project Director (PD)**, who may or may not be the project sponsor, located in the EPC Company office, shall be responsible for ensuring that communication and coordination are managed in accordance with the PCP and for communicating any additional requirements to the respective department personnel.

The EPC Program or Project Manager, located at the construction site, is responsible for managing the project communications and for the document management system. These responsibilities also include interfacing with the customer's site Program or Project Manager for all site-related issues that arise over the construction phase of the project.

The Document Controller, located in the EPC Company, is responsible for ensuring that all correspondence and communications are electronically documented and filed for easy retrieval of records. This includes taking those measures necessary for the protection of the documents and records from deterioration, damage, or loss. This person is also responsible for compiling the Project Document Plan, which is discussed in Chapter 4, and establishing a site Document Control Center with the assistance of the Program or Project Manager.

The EPC Construction Manager, located onsite in the Construction Office, will be responsible for managing the communications onsite for the EPC Project Management Team and is the focal point for interfacing with the customer's project team onsite, which includes the customer's Construction Manager.

Project Team Directory

Table 3-3 represents a standard example of contact information for all site managers identified in a normal communications management plan. The email addresses and phone numbers, if available, will be used for primary communication onsite and offsite during construction, which is especially critical in the event of a site emergency.

Table 3-3 Directory of Construction Site Project Management Team for Communication Management Plan

Title	Name	Email	Phone	Location
Customer Project Director				Customer Corporate Office
EPC Project Director				EPC Corporate Office
Customer Program or Project Manager				Construction Site Office
Customer Site Construction Manager				Construction Site Office
EPC Construction Manager				Construction Site Office
EPC Project or Program Manager				Construction Site Office
EPC Site Manager				Construction Site Office
EPC Site HSE Manager				Construction Site Office
Customer QA/QC Manager				Construction Site Office
EPC QA/QC Manager				Construction Site Office
Customer Safety Officer				Construction Site Office
EPC Document Controller				Construction Site Office
EPC Procurement Manager				Construction Site Office

Communication Matrix

The different locations for a complex construction project also require different communication requirements, which leads to a more complicated Communication Matrix for this type of project. An example of this type of Communication Matrix for a large, complex construction project with multiple offices is shown in Figure 3-6.

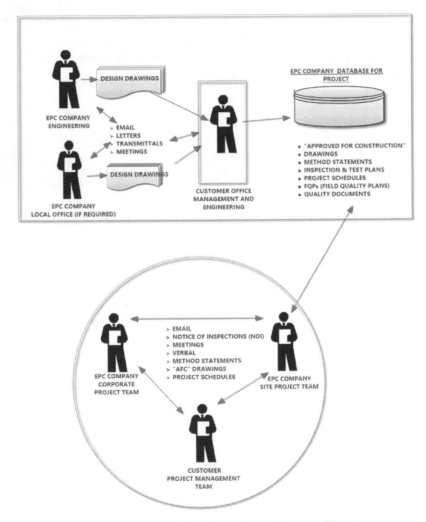

Combined Communication Matrix for a Typical Large, Complex Construction Project

Figure 3-6 Combined Communication Matrix for a Typical Large, Complex Construction Project

The critical item in a complex situation is that all communication between each group must be concise and clear, and that feedback is provided in the time frame required by the sender.

Communication Methods

The language to be utilized in all forms of communication for a large, complex construction project is usually specified in the project's contract and in various forms, as follows:

1. **E-mails**

   E-mails are used for daily communications among all of the parties as they relate to project activities. E-mails can also be used as a means for document transfer, Minutes of Meeting (MOM) transmittal, and/or for an official letter or transmittal with an authorized signature.

2. **Letters**

   Letters, which shall be limited to one subject only, are to be used for communicating contractual, schedule, and significant technical issues regarding contract execution and, for this reason, can only be transmitted with approval of the Project Director. Letters used to seek clarifications on non-technical issues, making proposals/approvals, conveying information related to all project matters, etc., can be initiated by lead personnel in either the customer's or EPC Contractor's corporate office, but they must have the Project Director's approval prior to transmittal.

3. **Memorandums**

   Memorandums are to be used for formal communication among the customer's or EPC Contractor's internal parties, which also includes all stakeholders.

4. **Drawing and Document Transmittals**

   Transmittals shall be used for submission of documents produced by the EPC Contractor as well as some routine administrative documents. Transmittals shall also be used by the customer as comments for notification of approval of documents and drawings.

5. **Minutes of Meetings**

   Minutes of Meetings (MOMs) conducted at the construction site, in local offices, or out of the country shall be recorded, distributed, reviewed, and approved by members of all relevant parties involved in the project prior to their submission to senior management.

6. **Verbal Conversations**

On construction projects, many small problems onsite can be solved with just a brief verbal discussion between the respective EPC Construction Manager and the customer's Construction Manager. This is supported by the Communication Plan so that the work onsite can be expedited effectively. After these problems are resolved, the EPC Construction Manager should record this conversation and the decision in a daily log for the project.

Communication Processes

Official Letters

If there is either a management or an engineering issue onsite that the EPC Contractor's Site Program or Project Manager cannot resolve with the customer through meetings or verbal communication, then this issue is sent to the EPC Contractor's local or home office. The time required from the moment this issue is known until the customer is officially notified is usually specified in the project's contract because it may result in a change order to the contract. This process is graphically portrayed in Figure 3-7.

This communication process can take about two or three days, but it permits the EPC Contractor's Project Director to retain control of all formal communication with the customer at the project management level for quick action, which usually involves contractual issues.

Meetings

There will be a number of meetings onsite on a weekly basis, and each one has its own purpose and order in which it will be conducted. These weekly onsite meetings for a typical large, complex project will be as follows:

1. **Weekly Progress Meeting** – This meeting will be a site management meeting with management from the EPC Contractor and customer; its purpose is to resolve any issues onsite that will affect either the project's progress or the health and safety of all personnel working onsite. This meeting will be chaired and recorded by the EPC Contractor with each MOM distributed to the senior management of both companies. It is imperative that meetings onsite be limited to one hour maximum, if possible, because the EPC Contractor's management should be onsite as much as possible.

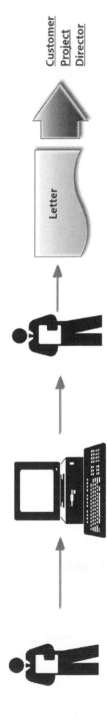

PROCESS FOR AN OFFICIAL LETTER TO BE ISSUED FOR A SITE MANAGEMENT OR ENGINEERING ISSUE

Customer Project Director

Letter

EPC Contractor Management Team Onsite Identifies a Site Engineering or Management Issue

Sends an Email to the Design Engineer or Project Director in the EPC Company Office

Letter Signed by EPC Contractor Project Director is Sent to the Customer Project Director Regarding the Issue and Requesting Them to Take Immediate Action for Correction.

Figure 3-7 Process Flowchart for an Official Letter to Resolve a Site Issue

2. **Weekly Interface Meeting** – This meeting will be an informal meeting (No MOM) with management and engineers from the EPC Contractor and customer for the purpose of improving coordination of the work activities being performed onsite by various contractors. The reason for not recording the minutes of this meeting is to keep this meeting short so that site management can spend more time onsite. The outcome should lead to expedited resolution of any immediate problems onsite and early identification of support requirements that each party may have in the forthcoming week.

3. **Weekly Schedule Meeting** – This meeting will be a formal meeting, which will involve updating the current spreadsheet for all critical work activities with input from the planners/schedulers (Project Control) from the EPC Contractor, EPC Contractors, and the customer for the purpose of aligning with the monthly Level 5 schedule. This updated project schedule spreadsheet alone is then formally submitted each week to the senior management of all companies, along with the site-specific three-week "look-ahead" Level 3 schedule. The expected outcome is a more accurate list of critical work activities that should assist the site management from the EPC Contractor and customer with improving the current project schedule.

4. **Daily Contractor Coordination Meeting** – This meeting is normally held onsite at a specified location with a representative from each of the contractors so that the EPC Construction Manager can coordinate all site work activities for the next day. This coordination includes items such as crane support, road access, work access, etc.

Reports

The reports for this project, which are contractual obligations, will be used to monitor the progress of work being performed onsite and the current status of the project's schedule with relation to specific milestones as well as to provide the overall Health & Safety condition of the work environment onsite. These reports, along with the party responsible for their compilation and submittal, will be as follows:

1. **Daily Progress Reports**
   - *Customer* – The customer's Project Engineer will be responsible for the compilation and production of a daily progress report, which will involve feedback from each member of the customer's onsite team.
   - *EPC* – The EPC Construction Manager will be responsible for the compilation and production of a daily progress report, which will

involve feedback from each contractor's staff onsite and include daily manpower totals for the EPC Company and all of its contractors.

2. **Weekly Progress Reports**
   - *Customer* – The customer's Project Engineer will be responsible for the compilation and production of a weekly progress report, which will involve feedback from each member of the customer's onsite team because their management along with the EPC management will be viewing this report.

3. **Schedule Updates**
   - *EPC* – The EPC Program or Project Manager will be responsible for the compilation and production of the following project schedule updates:
     - ➢ Weekly Level 3 Schedule Update
     - ➢ Monthly Level 5 Schedule Update

These schedule updates will involve feedback from the EPC Contractor's project control group onsite and the EPC Contractor's planning engineers onsite.

These reports will be sent on a timely basis to all of the site management for all companies and to the respective senior managers of each company. The intent is to ensure that all managers involved with executing and managing this project are working with the same information.

Notice of Inspection (NOI)

The Project Quality Plan requires that all areas of a work activity that are hold points and require the customer's engineer witness will be performed by providing a Notice of Inspection (NOI) to the respective customer's engineer 24 hours in advance of the time for inspection of that activity. The EPC Contractor has this responsibility on the construction site, and the customer's onsite Engineering staff will perform the required inspections to clear this hold point so work can continue. The reason behind the 24-hour notice is that it will provide the inspecting Engineer time to review all of the documents required for this inspection and ensure that the work is being performed according to the latest revised drawings.

3.2 Concluding Remarks

The key to the success of any project is effective communication between all stakeholders involved with managing and directing the project, which can be either a Project Director, Program or Project Manager, a Manager or a Project

Engineer. The one area on large, complex construction projects that tends to get out of control are meetings that keep the EPC site management tied up and off the construction site. If the project's progress starts to slide, the tendency of the customer is to call for more meetings, which, in most cases, leads to further project delays. The rule of thumb for site meetings should be a maximum of one hour and to ensure that the agenda is followed by all parties.

The intent of this Project Communication Plan is to provide a road map that the EPC Contractor, Program or Project Manager, and the project team will follow over the life of this project. A road map will help ensure success through an effective flow of clear, concise, and purposive information at every level of the project organization.

In the next chapter, Part III of the Project Management System, we will examine not only the Project Risk Plan but also the Document Plan and Resource Plan. All of these plans require concise and clear communication for continual updating over the life of the project to maintain their effectiveness for the project.

Chapter 4

Project Management System: Part III

4.1 Project Management System

In this chapter—Part III of the Project Management System (PMS)—we will examine not only the Project Risk Plan, but also the Document Plan and Resource Plan, which are also critical to the effective operation of the Project Management System.

4.2 Project Risk Management Plan (PRMP)

The determination of risk for a particular activity is an iterative process that starts with first analyzing what items can delay that activity or increase its operating costs. After this analysis is performed, the next step is to determine the probability and impact of these items individually using the matrix shown in Figure 2-2, which was discussed earlier, in Chapter 2. The best way to understand this process is to individualize this procedure through an example that most people have experienced in their lives, which is driving a car. Each person that decides to drive a car is aware of the risk of an accident but accepts this risk as possible with a 30% chance and 2 on the impact scale (see Figure 2-2) because they mitigate this risk through driver education and car insurance. As they gain experience driving, this risk moves from unlikely to remote unless they have an accident, when, in most cases, it now moves to likely (a 50% chance

Phase	Risk Level
Civil	80.00%
Mechanical	60.00%
Electrical	40.00%
I&C	16.00%

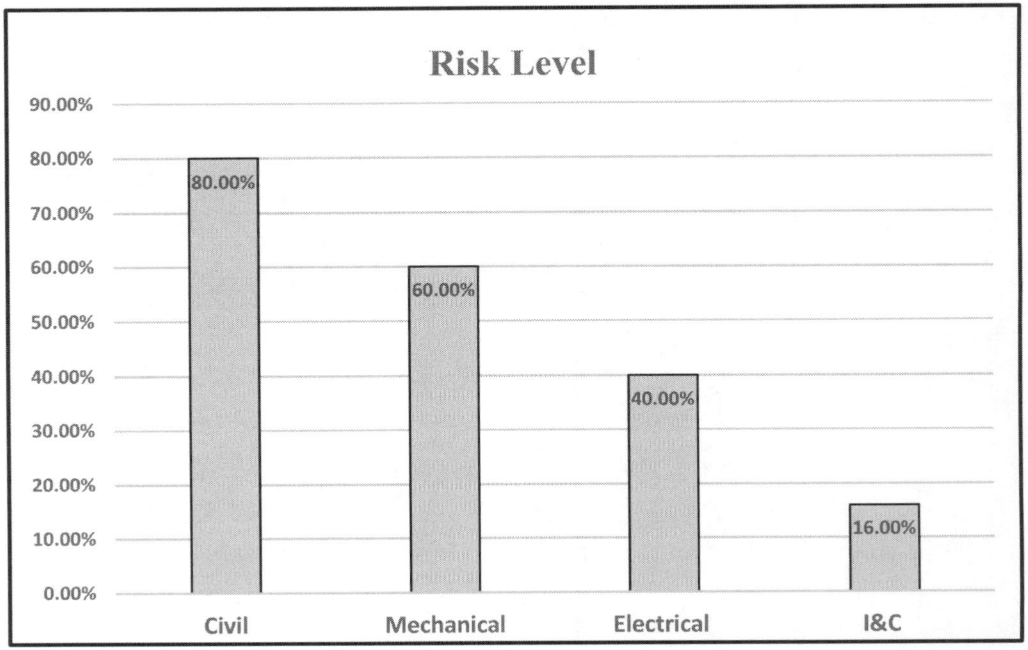

Figure 4-1 Map of Risk per Phase for a Large, Complex Construction Project

and 3 on the impact scale), and their car insurance premiums increase because the insurance company feels risk has increased. When developing the PRMP, this approach is used to analyze each activity—especially those on the critical path—as well as any additional internal and all external threats, such as heavy rains, flooding, fire, civil unrest, etc. This process involves the following steps:

Step 1 – Segregate the analysis into the four phases of the project—civil, electrical, mechanical, and I&C (Instrument and Controls). The standard risk levels for a large, complex project tend to look like the risk map example shown in Figure 4-1.

The reason that the civil phase has the highest level of risk—80%—is that it starts with a large area of land with many unknown risks that are not usually discovered until excavation begins. At the same time, civil work also has to contend with natural elements, such as storms, flooding, snow, etc., which can shut down the work for a period of time. The mechanical phase is the second highest—60%—because it usually starts after the civil phase has completed the required foundations for equipment and the necessary structures required for the piping. But at the same time, many problems do not arise until the installation of equipment or piping begins. The electrical and I&C phases are lower because many of their risks are known and usually mitigated through specific installation instructions, which reduces a majority of the risk associated with these activities. This risk mapping for all four phases of a construction project should be performed prior to the start of the project's execution phase. It will provide two important items to the EPC Program or Project Manager and Construction Manager, as follows:

- There is a much greater risk for cost overruns at the beginning of construction, and this risk continues until the civil work is almost completed.
- A greater amount of work must be focused on establishing the correct mitigation strategies for civil and mechanical work because they pose the greatest threat to the project if something goes wrong.

Step 2 – In this step, an analysis of the area where the construction activities will be performed must include reviewing the following documents:

- **Site Geotechnical Report** – This is a geophysical evaluation by a civil engineering company, which is experienced in producing very accurate and complete geotechnical analysis reports, of the surrounding surface area where the construction work is conducted. This analysis should include taking various soil core samples in the area where the large foundation will be erected, for the purpose of determining if the soil can support the loading that will be associated with each foundation. If this area is part of an existing development, this geotechnical report should not only establish the boundaries for the site, but it should also determine if there are any underground services present underneath the soil. The other item that can impact the civil work is whether this area has any underground water issues along with possible seismic activity, which would also change the design of the buildings. All of these items should be in this site geotechnical report.

- **Meteorological History** – This information is required to determine the probability of flooding, sudden torrential downpours, high winds, heavy snowfall, etc. This knowledge is critical to developing the correct mitigation strategy for the civil excavation and foundation work.

4.2.1 Risk Identification Key Input(s) and Risk Register

The key inputs and outputs for the risk identification process are shown in Figure 4-2.

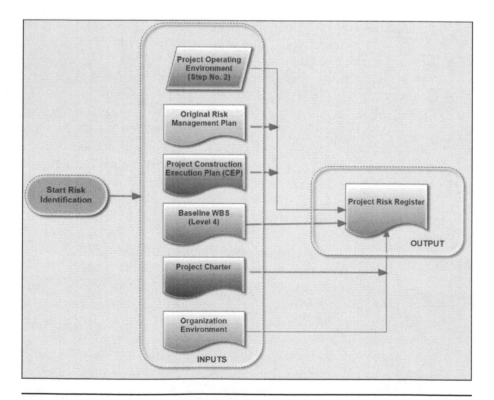

Figure 4-2   Inputs and Outputs for the Risk Identification Process

The project operating environment evaluation performed in Step 2, above, is critical because the theater where the project will be performed, which could be a domestic location, an international location, a heavily regulated defense or government facility etc., can have many inherent risks by its very nature. The WBS input, shown in Figure 4-2, should be at a minimum of Level 4 so that all of the possible "triggers" that initiate most risk events can also be identified during the risk identification process. The final product of this step is the actual Risk Register for this project, which is the backbone of the PRMP. An example of a Risk Register for a project is shown in Table 4-1.

When developing this Project Risk Register, it is critical that the WBS number of the associated activity or activities is included so that the Program or Project Manager can readily evaluate the impact on the project's schedule and budget if this risk is not mitigated. The EPC Construction Manager is the person who will be responsible to ensure that all mitigation measures are in place onsite and implemented if the risk is triggered by the event identified in the Risk Register.

Once the Risk Register is completed, which should be done before the start of construction activities onsite, the final step is for the EPC Program or Project Manager and Construction Manager to review, quantify, and then sort all risks so that the risks with the highest scores are at the top. The intent here is to effectively utilize the site resources, along with the various project resources, in mitigating those risks with the greatest impact on this project during its execution phase.

The next part of the PRMP is to establish the communication method for notification when a risk mitigation is initiated, which should be concise information and quick, because the EPC Program or Project Manager along with the EPC Company's management does not have time to spend on producing or reading reports. This method should be agreed upon by all parties within the EPC Company and clearly established in the PRMP. In addition to this communication, the following project documents should be updated once the mitigation work is in progress:

- Project Financial Register
- Project Construction Execution Plan
- Project's Level 5 Baseline CPM schedule (if there is a schedule impact)
- Project Risk Management Plan

During the construction of a large, complex project, threats will arise that are considered to be "unknown risks" and require immediate mitigation action to be followed. The process that should be followed for these "unknown risks" is shown in Figure 4-3.

The inputs shown are not only required to attempt to identify this unknown risk but also to identify it in relation to cost and time impact upon the project's

Table 4-1 Example of a Risk Register for a Large, Complex Construction Project

PROJECT NAME: LARGE COMPLEX CONSTRUCTION

RISK IDENTIFICATION			QUALITATIVE RATING				RISK RESPONSE		
RISK No.	RISK	CATEGORY	PROBABILITY	IMPACT	RISK SCORE	RANKING	RISK RESPONSE	TRIGGER	RISK OWNER
1	Typhoon and heavy rainstorms with lightning	Work Environment (WBS No. 1.1)	9	9	81	Almost certain	1. Monitor local weather forecast and plan activities that are indoor work, where possible 2. Increase site drainage to mitigate impact of flooding	Dark storm clouds and strong prevailing winds at site	EPC Contractor
2	Concrete Batching Plant cannot fulfill project concrete requirements	Resources (WBS No. 1.2.1)	5	7	35	Unlikely, but possible	1. Preplan all large concrete pour activities and provide Batching Plant advance notification of requirements 2. Check if 2nd Batching Plant is available	Many large foundations requiring pouring at the same time	EPC Contractor

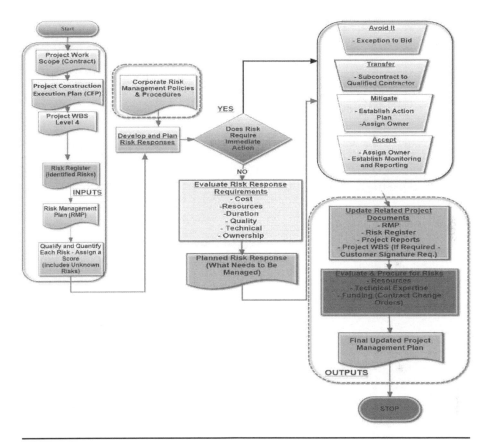

Figure 4-3   Mitigation Response Process for "Unknown Risks"

schedule. There are cases in which this risk may be something outside of the contractual work scope and require a contract change order to be signed by the customer before mitigation work can proceed, which typically happens in force majeure situations, such as flooding, social unrest, sink holes, etc.

4.2.2 Concluding Remarks

The PRMP is extremely important to the project, especially when it involves a large, complex construction project, because it is the site Program or Project Manager's game book when something goes wrong onsite. It is imperative that as the civil, mechanical, electrical, or I&C work onsite progresses and finishes, the project's Risk Register along with the PRMP are also updated. This update process will not only incorporate "lessons learned" but, at the same time, will keep the game book current until the game of construction onsite achieves a perfect score for all project deliverables.

4.3 Project Document Plan (PDP)

A large, complex construction project requires thousands of drawings, procedures, quality documents, and engineering documents, which must not only be produced but also stored correctly. An example of the magnitude of the number of drawings required for a typical large, complex construction is shown in Figure 4-4.

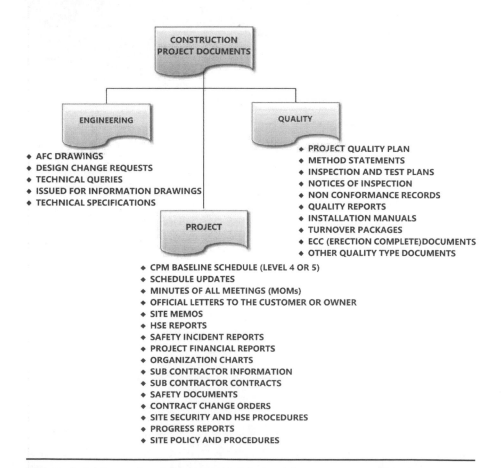

Figure 4-4   Example of Documents Required for Large, Complex Construction Projects

In many contracts, the production and transmittal of all project documents is usually specified to occur between the corporate office of the EPC Contractor and the customer because the construction site offices have not typically been erected at the time of contract signing. However, this should not preclude the EPC site Program or Project Manager from ensuring that a Document Control Center is established as one of the first offices onsite and is copied on all of these transmittals, which can be hard copies instead of electronic files. The PDP is critical in not only ensuring compliance with the contractual document requirements but also in how this Document Control Center will be established, its purpose, and how all documents will be stored onsite. This includes controlling all documents such as letters, memos, customer responses, etc., with an assigned number unique to the site Document Control Center so that the EPC Company knows what documents belong to the site.

The first part of the PDP should establish the mission of the EPC's site Document Control Center and how it will achieve that mission. This section of the PDP should also include a site organizational chart as shown in Figure 4-5.

EPC SITE DOCUMENT CONTROL ORGANIZATIONAL CHART

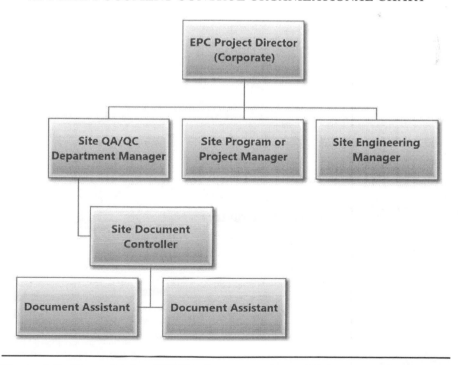

Figure 4-5 Example of an EPC Site Document Control Organizational Chart

The site Document Control Center is the responsibility of the site QA/QC Manager for the EPC Contractor because audits will be required on a specified time basis to ensure that all documents are being updated and being filed in the manner specified in the PDP. Another reason for the QA/QC Department to manage the Document Control Center is the issuance and control of all construction drawings for the EPC Contractor and its subcontractors. One of the most critical responsibilities will not only be the issuance of Approved for Construction (AFC) drawings to support the work onsite but also strict control over that issuance so that no work is being performed onsite to an outdated drawing. For example, if a civil drawing has been modified to show a different depth of excavation that is due to a foundation design change, then it will now be shown as Revision 2 in the Master Drawing List (MDL). When this newly revised drawing arrives onsite, then the EPC Document Control Center must first remove the old Revision 1 from both the EPC site Civil Engineer and the subcontractor before issuing the new Revision 2. If this is not done, then the wrong type of excavation could be performed, thus leading to expensive rework (which usually has a considerable cost and schedule impact upon the project). The best method for controlling this process is through an electronic spreadsheet (see Table 4-2).

Table 4-2 Spreadsheet for Tracking and Control of Approved for Construction (AFC) Drawings Onsite

Drawing No./ Revision	Discipline Type	MDL Revision	Does MDL Revision = Site Revision	Old Drawing Removed and New One Issued/Date
765C1048/Rev. 1	Civil	765C1048/Rev. 2	No	Yes/01/25/16

The reason that the site Engineering Manager and Program or Project Manager are involved is that their staff will be utilizing the Document Control Center for various documents to fulfill their assigned duties, and this information must be checked at times to ensure it is being updated (which is what these managers should do over the life of the construction project).

The other requirement of this PDP is a specification regarding how all of this information will be filed. Many EPC Companies prefer to follow the filing method employed by their corporate Document Control Center, which includes an electronic database. However, on many remote construction sites, the internet is usually either very poor or nonexistent, which means that the site Document Control Center must keep the files on a server located in the site office. Since these documents are critical to the EPC Company, the Program or

Project Manager should ensure that the site periodically downloads these documents onto the company's central server located in the corporate office because a fire or flooding of the office can destroy all of these files, which can prevent the project from being closed and the deliverables not being turned over to the client with missing documentation.

The last part of this PDP specifies the site printing requirements so that the Program or Project Manager can purchase the correct office printers, which should have the capacity to print both regular and legal size documents. If the project is outside the United States, then this capacity should be sizes A4 and A3 because this is the standard utilized in most foreign countries. When it comes to drawings, the EPC Company should give consideration to having at least one plotter in the Document Control Center so that full-size drawings can be printed for the Engineers and subcontractors, as these drawings are typically used onsite. After this procurement is completed, the next step is to make sure that these printers and plotter can be accessed from each person's work station in the office, which may require the service of the corporate information technology (IT) department to assist with this activity.

4.3.1 Concluding Remarks

On most projects, the tendency is to establish the site Document Control Center much later in the project because most companies do not have a PDP, which leads to many lost documents and improper cataloging of critical correspondence. This correspondence may not seem critical at the time, but later, if the Program or Project Manager decides to file a claim for an incident that occurred a few months ago, this correspondence becomes critical. The other area where documents get lost is when a company on a large, complex construction project does not keep everything on one central server, but rather leaves this to everyone having documents on their computers. If a computer crashes, which they do, then many critical documents, along with important emails, can be lost unless that person was backing them up on an external flash drive. The PDP is critical for these reasons, and trying to establish a document control center when the room is filled up with thousands of documents, erection manuals, drawings, etc., only leads to more chaos for everyone, and usually at the wrong time.

4.4 Project Staffing/Resource Plan

When a person is hungry, he or she has a need for food, and after eating, this need is fulfilled. This is the approach that will be followed to develop the Project

Staffing/Resource Plan (PSRP) for the project because each phase has different needs that must be fulfilled during that phase. Resources should be defined as both equipment and the subcontractors required to fulfill that need because a subcontractor without equipment cannot perform the work. The phases we will examine and use for this development are as follows:

1. **Site Establishment** – This phase involves the establishment of the perimeter fence, access roads, site offices, and the required services to support both the customer and EPC Contractor site teams.
2. **Site Mobilization** – This phase involves mobilizing the EPC contactor's project team and its subcontractors to the site for the start of work.
3. **Civil Work** – This phase involves the actual start of construction work onsite, which is primarily civil.
4. **Electrical Work** – This phase involves the actual start of electrical work onsite, which can, depending upon the project, have its beginning later in the Civil Work phase.
5. **Mechanical Work** – This phase involves the actual start of mechanical work onsite, which can, if there are early underground piping requirements, have its beginning later in the Civil Work phase.
6. **I&C Work** – This phase involves the actual start of I&C work onsite, which usually starts after the Electrical Work phase has begun.
7. **Commissioning** – This phase involves the turnover of the deliverables from civil and their subsequent initial energization or operation as specified for each deliverable in the contract.
8. **Site Demobilization** – This phase involves the turnover of the operating deliverables from the EPC Contractor to the customer. After this is done, then the EPC Contractor starts removing its remaining subcontractors from the site along with their site project team.

These eight phases clearly show how their needs are different, which is why the PSRP must not just be an organizational chart to fill with names and faces, but rather a dynamic placement of resources as they are needed to make the project a success.

After the tables are completed for each phase of the large, complex construction project, the following steps are required to complete the PSRP:

a. Incorporate all of the tables, discussed above, into the plan for review and comment by the site Program or Project Manager and the Site Construction Manager for the EPC Company.
b. Color code each position in the Site Organizational Chart with the same color from the table in Step Number 1 as it applies to that position. If

there any positions that need to be added, add them and color code each one accordingly. This will ensure that the PSRP is in alignment with the Project CEP and the Level 5 Baseline CPM Schedule for the project.

c. Build a Site Organizational Chart for all of the subcontractors and perform the same color coding for each one as was done in Step Number 2. The timing of a subcontractor on a site is critical to prevent that subcontractor from experiencing waiting time with their personnel and equipment, which costs money and can lead to a claim against the EPC Contractor.

d. In this final step, it is critical to establish the goals for each site group— Engineering, QA/QC, HSE, Document Control, and Projects. After these goals are accepted, clearly define the roles and responsibilities for each position within these five groups. This will lead to a more effective site team because all players understand their mission and how it fits into the overall goals of the project.

In many projects, the customer usually has to review and approve subcontractors that the EPC Contractor chooses for the construction work. If a subcontractor has not received that approval, then it should be noted when the subcontractor's organizational chart is received and presented to the customer.

4.4.1 Site Establishment

The first thing the Program or Project Manager will require is the general arrangement drawing, approved by the Engineering Department of the EPC Contractor and the customer, which will establish the construction site. The site establishment work starts after the Site Permit is approved and received by the EPC Contractor. In order to develop the PSRP for this phase, a table of needs is built, followed by the items required for their fulfillment, which is shown in Table 4-3.

An examination of Table 4-3 shows how the satisfaction of the first two needs in this phase also satisfies the needs for most of the remaining activities. For example, the Qualified Survey Team in the first activity is utilized for the remainder of the activities, so their need is not listed because it is fulfilled. The other item to remember is that if the two qualified Civil Subcontractors utilized in this phase can be utilized for the future Civil Work onsite, then the project will save money and time on additional mobilization costs. If the Program or Project Manager knows the duration of these activities, which should be established in the Project Construction Execution Plan, a time frame for when the staff and resources are required during the Site Establishment Phase can be established.

Table 4-3 The Project Staffing/Resource Plan (PSRP) for Site Establishment

Activity	Needs	Staffing Required	Resources Required
Establish perimeter fence around site w/gate	• Fence Material • Steel Posts • Qualified Subcontractor • Concrete • Gates (2) • Site surveyor	• EPC Site Civil Engineer • EPC HSE Engineer • EPC Qualified Civil subcontractor • EPC Security Personnel • Qualified Survey Team	• Boring machine for post holes • Concrete for post holes • 2 large gates that slide on tracks • Fencing material and clips
Establish access roads around site and site drainage	• Subcontractor experienced in road building • Equipment for soil removal, grading, backfilling and compaction • Pre-cast concrete structures for site drainage and piping • AFC Civil drawings for the road and drainage ditches	• EPC Qualified Civil subcontractor with experience in building roads • EPC Civil Design Engineer access during process for support • EPC Construction Manager • 2nd EPC Site Civil Engineer	• Approved material for the new roads such as crushed rock, soil mixture, etc. • Excavators, Dump Trucks, Backhoes, Compactors, and Graders (Machinery) • Area for dumping of removed material
Establish location of site offices and install them	• AFC Civil, Mechanical, and Electrical drawings for office locations • Approved drawings for Offices • Mechanical potable water piping • Plumbing fixtures • Office furniture • Electrical Cable, Conduits, etc. • Fire Detection System • Portable Generator Sets	• EPC Electrical Engineer • EPC Mechanical Engineer • EPC Qualified Electrical subcontractor • EPC Qualified Mechanical subcontractor • Office Cabin supplier • Generator Set supplier • Structural Steel and Roofing supplier	• Office Cabins • Conduit for the high-voltage generator cables. • Approved material for the site offices and generator sets, such as crushed rock, soil mixture, etc. • Concrete and formwork for Generator Set Enclosure • Mechanical and Electrical service material, such as PVC pipes of various sizes, conduit of various sizes, electrical panels, etc. • Office equipment
Acceptance of Site Offices and their layout by the Customer	• QA/QC documents • Inspection forms	• EPC Site QA/QC Manager or Engineer. • Customer's QA/QC Manager	• EPC QA/QC inspection forms and punch lists

4.4.2 Site Mobilization and Remaining Phases

The intent of this phase is not only to bring all of the EPC Contractor staff to the site, but, at the same time, to also establish the Document Control Center and the communication systems that will support the staff over the construction period of the project. Table 4-4 demonstrates the needs to be filled during this phase.

The one thing to remember is that because some EPC staffing was added in the Site Establishment phase of this project, staff mobilization was reduced because these needs were already addressed. Another item to note is that in this phase, there should be preplanning for the next phase, Civil Work, by mobilizing the required EPC Qualified Civil Subcontractor and Qualified Electrical (Industrial) Subcontractor early so that they will be fully manned and ready to start

Table 4-4 The Project Staffing/Resource Plan for Site Mobilization

Activity	Needs	Staffing Required	Resources Required
Establish Document Control Center	• Document Controller • Shelving Units and Document Storage Cabinets • Industrial Printing Stations (2) • Server and Internet Network • Document Control Assistant • Work stations (4) • Office Supplies • Plotter	• EPC Document Controller • EPC Document Assistant • EPC IT Corporate person (temporary until work is completed)	• Office Supplies • Two Printing Stations • Qualified Internet Company for support and wiring • Plotter • Four Work Stations • Shelving Units and Document Storage Cabinets
Mobilize EPC Contractor to Site	• Site Program or Project Manager • Admin. Assistant • Qualified Civil Subcontractor • Qualified Electrical (Industrial) Subcontractor • QA/QC Manager • Civil QA/QC Engineer • Electrical QA/QC Engineer • Civil Inspectors (2) • HSE Manager • HSE staff (4) • Site Transportation • Site Engineering Manager	• EPC Site Program or Project Manager • EPC Admin. Assistant • EPC Qualified Civil Subcontractor • EPC Qualified Electrical (Industrial) Subcontractor • EPC QA/QC Manager • EPC Civil QA/QC Engineer • EPC Electrical QA/QC Engineer • EPC Civil Inspectors (2) • EPC HSE Manager • EPC HSE staff (4) • EPC Site Engineering Manager	

Table 4-5 The Project Staffing/Resource Plan (PSRP) for Civil,
Electrical, Mechanical, and I&C Phases

Phase	Activity	Needs	Staff Required	Resources Required
Civil	• Civil Work	• Qualified Civil Construction Subcontractor • Qualified Civil Steel Erector • Cranes	• EPC Qualified Civil Construction Subcontractor (No. 2) • EPC Qualified Civil Steel Erector	• Cranes
Electrical	• Installation of Grounding and All Electrical Components, Including Wiring	• Qualified Earthing Cable Installation Subcontractor • Earthing Cable • Qualified Electrical Subcontractor (No. 2) • Electrical Cable (Various Types) • Qualified Electrical Terminators • Electrical Panels (Various Types) • Site Electrical Engineers (2) • Site Electrical QA/QC Inspector (2)	• EPC Qualified Earthing Cable Installation Contractor • EPC Qualified Earthing Cable Installation Subcontractor • EPC Qualified Electrical Terminators • EPC Site Electrical Engineers (2) • EPC Site Electrical QA/QC Inspector (2)	• Earthing Cable • Electrical Cable (Various Types) • Electrical Cable (Various Types)
Mechanical	• Installation of All Equipment, Piping, Firefighting Systems, Tanks, and HVAC in Various Areas	• Qualified Piping Subcontractor • Qualified Mechanical Subcontractor (Equipment) • Qualified Mechanical HVAC Subcontractor • Mechanical Piping Engineer • Mechanical HVAC Engineer • Mechanical Equipment Engineer • Mechanical QA/QC Inspectors (2) • HVAC Equipment and Ducting • Mechanical Equipment • Mechanical Piping of Various Sizes and Materials	• EPC Qualified Piping Subcontractor • EPC Qualified Mechanical Subcontractor (Equipment) • EPC Qualified Mechanical HVAC Subcontractor • EPC Mechanical Piping Engineer • EPC Mechanical HVAC Engineer • EPC Mechanical Equipment Engineer • EPC Mechanical QA/QC Inspectors (2)	• HVAC Equipment and Ducting • Mechanical Equipment • Mechanical Piping of Various Sizes and Materials
I&C	• Instrumentation Installation and Connection	• Various Types of Instruments • Control Type Cables		• Various Types of Instruments • Control Type Cables

the Civil Work as scheduled. After the construction work starts onsite, the EPC Contractor staffing and resource needs should be fulfilled unless there is scope change, which would require a signed contract change order from the customer.

The next four phases of construction—Civil Work, Electrical Work, Mechanical Work, and I&C Work—primarily overlap one another at some juncture in the construction work. The EPC Construction Manager has to ensure that the equipment and human resources each subcontractor brings to the site match the requirements laid out in the Project Construction Execution Plan (CEP). Because these four phases overlap and their needs are much smaller, they can be combined into one table, as shown in Table 4-5.

The Commissioning Phase staffing requirements are primarily the EPC Commissioning Engineer and staff of Commissioning Engineers, which must also include an HSE Commissioning Engineer and a QA/QC Commissioning Engineer. If there is any support required, then one or more of the construction subcontractors is kept onsite with a small crew for this purpose.

After the Commissioning Phase is completed, then the deliverables for the project are handed over to the customer or Owner officially, which is when the Demobilization Phase commences, and it does not finish until all contractors, Site Engineers, and the Site Project or Program Manager have left the site.

4.4.3 Concluding Remarks

For most large, complex construction projects, the tendency is to simply provide an organizational chart for the Project Staffing/Resource Plan (PSRP) with a small introduction at the beginning to describe its goals for the project. However, as the discussion above shows, the PSRP can also be a valuable tool to effectively place staff and other resources as they are needed and to pre-plan the work to reduce cost and delays to the project.

In the next chapter, we will see how the Work Management System takes the inputs from the Project Management System (PMS), discussed in these last three chapters, and uses these inputs to effectively coordinate and manage all of the work onsite.

Chapter 5

Work Management System

5.1 Work Management System

In a large, complex construction project, the one critical system used during the execution of the work is the Work Management System. The goals of the Work Management System are to:

1. Manage the work of all EPC Contractors so that their milestone dates are in alignment with those for the project and the Level 5 Baseline schedule
2. Provide early warning to the EPC Program or Project Manager that an area of work is starting to fall behind schedule so that corrective action can be implemented immediately
3. Ensure that all of the costs involved with the execution of the work are aligned with those established in the project's budget
4. Provide the EPC Construction Manager with assurance that the work being performed during each phase of construction is following the project's Construction Execution Plan (CEP)
5. Ensure that the work being performed is in accordance with all specified EPC HSE guidelines and contractual specifications
6. Monitor and report information weekly to EPC management, as well as the customer, regarding the progress of the construction work onsite and any emerging issues that require management's support for resolution

During the execution of the construction project, the EPC Construction Manager has the responsibility of establishing the Work Management System, which should be done at the time of Site Establishment because contractors are already onsite at that time. Because the site construction work changes over the life of the project, specific items employed for one group of activities will not be the same for another group of activities.

5.1.1 Civil Construction Work

The civil construction work on a large, complex construction work typically comprises three phases (see Figure 5-1).

For example, in the first phase of Civil Work, the majority of the work activities involve soil improvement and excavation in various areas for the future establishment of foundations, which are different depending upon the structure that will be built in that area. This type of work is primarily done with various types of machines and only a very small number of workers. One of the ways to manage this work is shown in Table 5-1.

The first thing the EPC Construction Manager should do is view these foundations in the Project CEP and look at the analysis provided to determine what types of resources and equipment to use. Then, he or she should compile a concise work scope for the excavation bid specification that the EPC's Finance Department will require to submit to at least three contractors for this work. After the contractor for this excavation Civil Work is selected, the first step is to have a meeting onsite with the contractor's Project Manager and team so that they can examine the area where this work will be performed. The next step is to receive a Level 5 schedule from the contractor for this work along with its execution plan so that these documents can be reviewed and approved prior to the contractor mobilizing to the construction site. In order to establish a daily monitoring schedule, it is critical to provide this contractor with a Baseline daily excavation chart, which is shown in Figure 5-2.

This chart is based on a continual seven-day-a-week work schedule, which may not be the schedule that the contractor will work. But what is important is that the total amount of soil excavated each week matches the weekly total shown in this chart. In order to confirm that this is occurring, the EPC Construction Manager must have a brief weekly meeting with the Contractor's Project or Site Manager and view their current progress chart to confirm that it is in alignment with the EPC Company's expectations for this work. The other critical item is that the EPC Construction Manager must go out to the site occasionally and observe the excavation work, which should include counting the number of trucks moving in and out of each foundation site per hour. This will assist the EPC Construction Manager in determining if the weekly report provided by the Civil Contractor is accurate and if there are any inherent delays

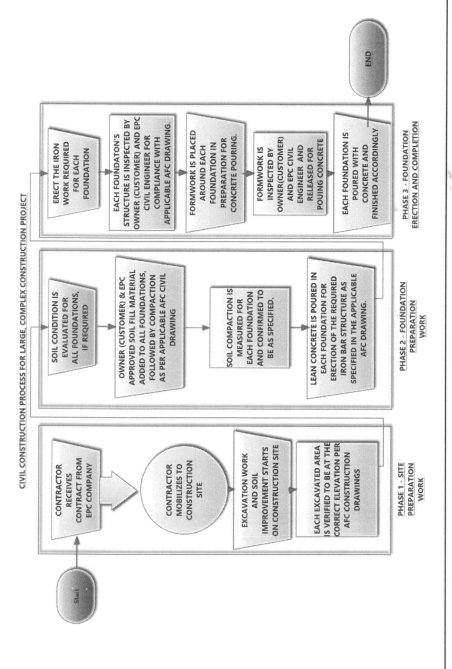

Figure 5-1   Three Phases of the Site Civil Construction Work

Table 5-1 Excavation Analysis for Management Purposes

WBS No.	Activity	Amount to Be Excavated	Planned Duration	Scheduled Completion Date
1.1.2.3	Excavate area for Foundation "A-5"	19,200 M³	30 days	30 June 2018
1.1.2.4	Excavate area for Foundation "A-2"	18,500 M³	45 days	15 July 2018
1.1.2.5	Soil Improvement for Foundation "A-3"	No excavation at this time until Soil Improvement is completed	15 days	15 June 2018
1.1.2.6	Soil Improvement for Foundation "A-4"	No excavation at this time until Soil Improvement is completed	15 days	15 June 2018

resulting from the methodology employed. For example, during the excavation of a large area on one construction site, this author noted that each truck was wasting 20 minutes just backing up into position for loading. When the contractor was told about this and how it could be corrected by establishing a circular route in the area, the 20 minutes for each truck went down to 5 minutes, and because there were about 100 loads per day being performed on a 10-hour basis, this allowed the contractor to increase their material removal rate. This increase also permitted this activity to get back on track and finish as scheduled.

Soil improvement is also usually identified during the site survey, but it requires a company specialized in performing this type of work. This requirement usually means that the EPC Construction Manager must carefully and specifically write a very concise technical scope, which will require assistance from the Civil Engineering Team, so that the company's Finance Department provides the bidders with a detailed work scope that must include all necessary testing requirements to confirm that the work is done in accordance with the required civil specifications. Each bidder must also provide a very detailed Level 5 schedule for the work being performed, which must be in alignment with the project's Level 5 Baseline CPM schedule that is supplied in the bid. After the contractor is selected, its Site Project Manager must contact the EPC Construction Manager to establish a prioritized list of the area that requires the soil improvement and all mobilization requirements prior to the contractor's arrival onsite. The tendency on most large, complex construction projects is to permit the existing civil contractor to subcontract this work, but if the work starts to fall behind schedule, the EPC Construction Manager tends to be informed late in the process. The late arrival of this critical information usually leaves little chance of schedule recovery and the possibility of large cost

WBS No.	Activity	Amount to Be Excavated	Planned Duration	Scheduled Completion Date
1.1.2.3	**Excavate area for Foundation "A-5"**	19,200 M³	30 Days	30-Jun-18
1.1.2.4	**Excavate area for Foundation "A-2"**	18,500 M³	45 Days	15-Jul-18

Day	1	2	3	4	5	6	7	8	9	10	11	12	13	14	15	16	17	18	19	20	21	22	23	24	25	26	27	28	29	30
Foundation A-5	640	640	640	640	640	640	640	640	640	640	640	640	640	640	640	640	640	640	640	640	640	640	640	640	640	640	640	640	640	640
Foundation A-2	411	411	411	411	411	411	411	411	411	411	411	411	411	411	411	411	411	411	411	411	411	411	411	411	411	411	411	411	411	411

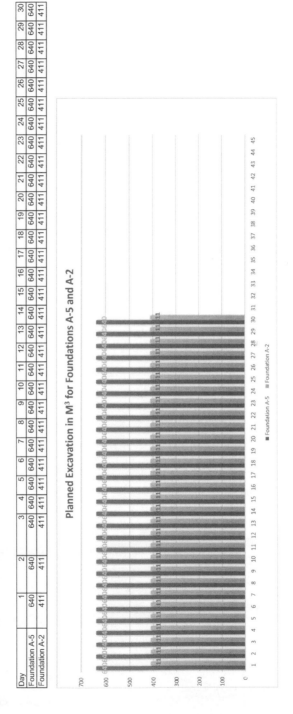

Planned Excavation in M³ for Foundations A-5 and A-2

■ Foundation A-5　■ Foundation A-2

Figure 5-2 Chart Showing Planned Excavation for Foundations A-5 and A-2

overruns for the project's budget to attempt any type of recovery. After the soil improvement contractor has arrived onsite and has started work, the EPC Construction Manager and the contractor's Site Project Manager must establish a concise and clear method to track the progress, which will be included in their weekly report to the EPC Company. At the same time, the EPC Construction Manager must also ensure that the current civil contractor is not waiting for the soil improvement contractor to finish its work, which will require close coordination between both contractors with the work they are performing. A simple example of this soil improvement process is shown in the steps below to illustrate the complexity of this work.

Step No. 1 – The site survey, as shown in Table 5-1, reveals that two of the existing foundation areas onsite must have spun piles installed to improve the current soil load-bearing condition, which must be done prior to excavation.

Step No. 2 – The EPC Construction Manager, along with the Civil Engineering Department, establishes that each of these two areas will require 100 piles driven to a specific depth, which is included in the bid specification for this work.

Step No. 3 – A reputable soil improvement contractor is awarded the contract and provides the EPC Construction Manager and Program or Project Manager with a schedule that shows completion of their work in accordance with the project's Baseline Level 5 CPM schedule. The soil improvement contractor also provides its mobilization program showing how it plans to bring four pile-driving machines onsite so that the team can work on two of the three areas at one time.

Step No. 4 – The soil improvement contractor has to first drive test piles in the two areas to confirm that the specified depth will provide the required load-bearing for the foundation to be erected.

Step No. 5 – After the testing in Step No. 4, then the pile installation work can commence, and the EPC Contractor along with the project team's Civil Engineer must continually monitor this work on a daily basis so that it is completed within the 15 days specified in Table 3-1. The best way to manage the progress is to simply count the number of piles being installed each day, and if it is insufficient, the pile installation contractor's Site Manager should be contacted immediately for action.

Step No. 6 – After the piles are installed, the work changes to cutting each of the piles to a specific elevation and excavating the area around the piled area for soil preparation. It is at this point that these two areas are turned over to the excavation contractor for the excavation, which usually requires the EPC Construction Manager to establish another separate schedule for this work so that it can be managed effectively.

After each foundation site has been excavated to the correct level, then the next phase begins—preparing the site for the erection of the foundation. The Approved for Construction (AFC) drawing for each foundation will specify what type of soil mix must first be added as well as the desired compaction for proper load distribution once the foundation is erected. The other requirement that must be met before this work begins is that the customer's Civil Engineers must also approve the soil mix because it will be part of the project's deliverables once the foundation is completed.

When the soil mix backfilling work begins, along with the subsequent compaction, the EPC Construction Manager should also establish the work scope for the third phase of the Civil Work (see Figure 3-1), which will require working with both the site and design Civil Engineers. The Project Construction Execution Plan (CEP) should have the machinery and personnel requirements, but the detailed erection of the iron work required for the foundation also needs to be added to the workscope for the bid proposal. In order to improve the schedule, it is essential to have another Civil contractor erecting the iron work foundation at the same time that the civil contractor is finishing all of the foundation backfilling and compaction work for each foundation (which includes pouring the lean concrete, once the compaction is accepted). This iron work will require a large amount of both skilled and semi-skilled workers. Therefore, the Civil foundation erection contractor, prior to mobilizing to the site, must submit to the EPC Program or Project Manager its personnel mobilization chart, which should look similar to the chart shown in Figure 5-3.

The EPC Construction Manager needs to review the amounts per month so that the contractor has enough people to complete all of this work as scheduled, to ensure that it is in alignment with the Level 5 Baseline CPM schedule for the project. The process for the erection of the iron frame of a foundation, as shown in Figure 5-4, clearly shows that there are four basic items that the EPC Construction Manager needs to focus on to ensure that each foundation is completed in time.

These four items are as follows:

1. **Iron Bar or Rebar** – There always has to be a sufficient supply of the right diameter of rebar onsite until all of the foundation has been completed.
2. **Bending Machines and Operators** – The contractor must have enough bending machines and qualified operators to ensure that this choke point does not lead to workers standing around waiting for rebar to install.

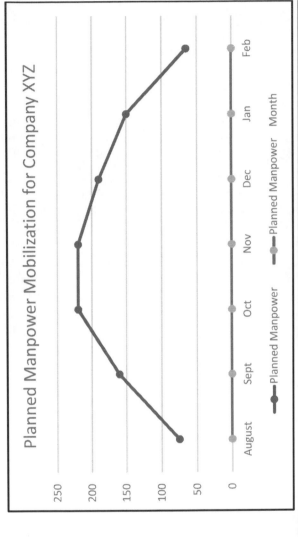

| Planned Manpower | 75 | 160 | 220 | 220 | 190 | 150 | 65 |
| Month | August | Sept | Oct | Nov | Dec | Jan | Feb |

Figure 5-3 Chart for a Civil Contractor's Personnel Mobilization on a Construction Site

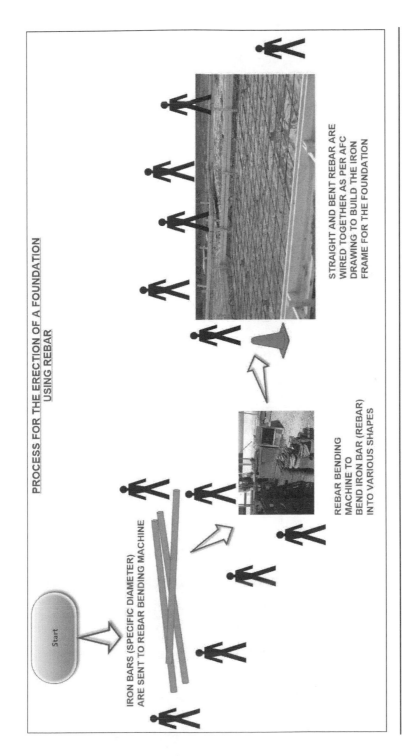

Figure 5-4 Process for the Erection of a Foundation Iron Structure

FOUNDATION TO BE POURED	CONCRETE QUOTE
Foundation "A-5"	2500 M^3
Foundation "A-2"	2000 M^3
Foundation "A-3"	1500 M^3
Foundation "A-4"	1450 M^3
Total	**7450 M^3**

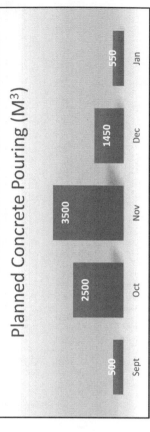

Month	Sept	Oct	Nov	Dec	Jan
Planned Concrete Pouring (M^3)	500	2500	3500	1450	550

Figure 5-5   Planned Concrete Pouring Schedule from a Civil Contractor

3. **Discipline Interface** – The EPC Construction Manager must contact the electrical and mechanical engineering managers to confirm that there is no cable, conduit, or piping that must be installed during the erection of each foundation. The site EPC Civil Engineer must also confirm that there are no embedded items that have to be affixed to the iron work. If any of these items are required, then action has to be taken immediately to incorporate this item into the erection process.

4. **Location and Transportation** – The contractor will need to bring enough mobile cranes to facilitate expediting the unloading of materials in the rebar bending yard and for each foundation being erected. The main focus here is that materials should be brought to the workers, and that the workers are not allowed to leave the work area for the procurement of material from the rebar bending yard. The EPC Manager and the site project team must monitor this work on a daily basis to ensure that this process is followed, to prevent delays in the erection process.

After the erection of each foundation, an inspection must be performed and approval received from both the EPC Civil inspector and the customer's Civil inspector. After this approval is received, the iron structure can be prepared for the pouring of concrete through the installation of formwork.

After the total amount of concrete has been established for each of the foundations, the civil contractor should provide a pouring schedule (depicted in Figure 5-5) for the amount of concrete that it plans to pour each month to the EPC Construction Manager.

The EPC Construction Manager uses this information to monitor the civil contractor's progress on all of the foundations so that each one is completed as scheduled. The civil contractor must provide weekly updates on this pouring schedule and notify the EPC Construction Manager if there are any delays. The other item that must be monitored by the EPC site Civil Engineers is the curing strength of the concrete for each foundation, which is performed by taking a series of samples of the concrete at the time of pour, which is called a "cube test" because these cubes will be checked for their compressive strength after 7 days and 28 days of curing. After this test produces the required results, the particular foundation can be released for erection of the required steel structure or specific equipment.

5.1.2 Electrical Construction Work

This area of work on most large, complex construction projects is typically underestimated, which leads to electrical construction work being started late when some of it should actually start at the time of the erection of the various foundations. In order to prevent this problem, the EPC Construction Manager should consult the EPC Electrical Engineering Department to confirm that none of these foundations required embedded conduits or an earthing (grounding) system to be installed in the rebar structure. The failure to perform this simple check can lead to a large amount of rework, which can have a serious impact on the project's budget and schedule. The EPC's site electrical engineer should be brought to the site at the start of the Civil Work so that all associated electrical drawings can be reviewed along with the Project's CEP to establish a clear scope of the electrical work at all construction phases of the project. The EPC's Construction Manager and the Program or Project Manager can then take steps to ensure that the Finance Department can solicit bids early so that the required Electrical contractors will start work as planned. On a large, complex construction project, the management of the electrical work has to be conducted in response to the location and nature of the electrical work being performed onsite. These particular situations and the best way to manage them are as follows:

1. **Underground Electrical Earthing (Grounding) System** – The best way to manage this work is to establish a grid system with the electrical contractor, as shown in Figure 5-6.

   Another way is to have the EPC site electrical engineer evaluate for each area the length of the tinned copper cable that needs to be installed and have the contractor update the information in Table 5-2 until all of the work is complete.

2. **Embedded Conduit** – In some buildings, there is embedded conduit within the foundation for the purpose of cables that bring power to various types of equipment or communication cables. In addition to foundations, conduit can also be installed in the walls of buildings and in the floor, for electrical connections. The best way to manage progress on the installation of conduit is to have the EPC Site Electrical Engineer conduct a daily inspection of all areas and mark up the drawing for each area to note what has been completed. The other item for foundations is to establish a date for the concrete pour and provide the date to the electric contractor so that they can ensure their work is completed in time for the release to pour concrete.

3. **Electrical Cable Tray Installation** – After civil has completed its work and has turned over the various areas to the EPC Site Electrical Team for its work, the work of installing cable trays and support in these areas for cable installation begins. The first thing is for the EPC Site Electrical Engineer to look at the Baseline Level 5 construction schedule

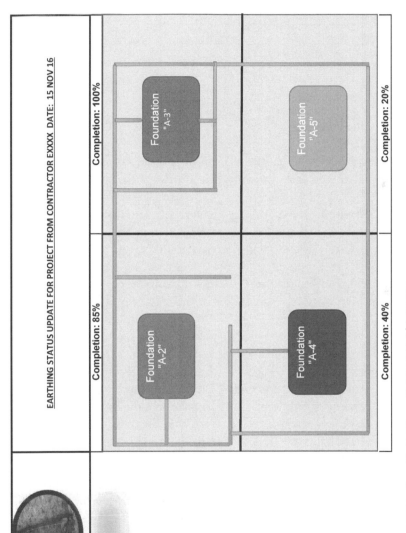

Figure 5-6 Grid System for a Construction Underground Earthing System

Table 5-2 Underground Earthing Cable Installation Progress

WBS No.	Location	Actual/Planned (Yards)	Percentage	Date
2.1.5.1	A-2 Foundation	1700/2000	85%	15 NOV 18
2.1.5.2	A-3 Foundation	2100/2100	100%	15 NOV 18
2.1.5.3	A-4 Foundation	720/1800	40%	15 NOV 18
2.1.5.4	A-5 Foundation	350/1750	20%	15 NOV 18

to determine what areas should be completed first by the Electrical contractor. The best way to manage this work is for the Electrical worker to provide a weekly update on the number of feet of conduit that have been installed, and if they are meeting the established targets (shown in Figure 5-7).

The amount of estimated time for this type of work should be in the Project's CEP and should be used to not only establish the scope of this work but also to evaluate if the electrical contractor has the ability to finish this work as scheduled.

4. **Electrical Cable Installation** – After the cable trays are installed, the next step is to start the installation of the electrical and control cables, which has to be a controlled process because all cables must be installed without any damage to their outer sheath. The condition of each pulled cable will be verified electrically once it has reached its destination and prior to its connection in a panel. The most important item is for the EPC Construction Manager and Electrical contractor to review the Level 5 Baseline schedule not only to prioritize the area where cable to needs to be pulled first but also to ensure that the work will be performed in strict alignment with the established durations and completion dates specified. This review will also establish how many feet of cable the contractor needs to be pulled each week to maintain this alignment (shown in Table 5-3).

5. **Electrical Termination in Panels** – After all of the cables for one panel have been pulled, the termination of each cable can begin. It is critical that the contractor provides qualification certificates for the personnel performing this work so that the termination work is performed in accordance with applicable specifications and quality standards. Because this is critical work that must be done manually, the progress can become slow, and the completion dates start to slide, which is why the Weekly Progress Reports shown in Table 5-4 are critical for the EPC Construction Manager, along with reports of any delays.

WEEKLY CABLE INSTALLATION PROGRESS REPORT

Day No.: **5**

Activity: Installing Cable Trays

Contractor: D X Y Electical

Start Date: 8-Nov-18

WBS No.	Site Location	Amount to Be Installed	Critical Path	Planned Amt/Day (FT)	Actual Amt./Day (FT)	Daily Progress	Total Amt Installed
2.1.6.1	Cable Cellar of Electrical Substation	1400 feet	X	70	60	85.7%	0
2.1.6.2	Utilities Room for Building 1A	2400 feet		80	75	93.8%	0
2.1.6.3	Roof for Building 1B	1200 feet		60	60	**100.0%**	0
2.1.6.4	Switchyard Cable Trenches	4500 feet	X	100	95	95.0%	0

*red- slipping

Figure 5-7 Cable Tray Installation Progress Report Example

Table 5-3 Cable Pulling Progress Weekly Report

WBS No.	Location	Weekly Actual/Planned (Yards)	Percentage	Week Ending Date
2.1.7.1	Cable cellar of Electrical Substation	1200/1500	80%	10 DEC 18
2.1.7.2	Utilities room for Building 1A	1450/1600	90.6%	10 DEC 18
2.1.7.3	Roof for Building 1B	850/850	100%	10 DEC 18
2.1.7.4	Switchyard cable trenches	2250/2500	90%	10 DEC 18

Table 5-4 Cable Termination Weekly Progress Report Example

WBS No.	Location	Weekly Actual/Planned (Terminations)	Percentage	Week Ending Date
2.1.8.1	Relay Panels in Electrical Substation	675/1800	37.5%	24 DEC 18
2.1.8.2	MCC Panels for Building 1A	300/1350	20.2%	24 DEC 18
2.1.8.3	Chiller Units on Roof for Building 1B	240/1000	24%	24 DEC 18
2.1.8.4	Switchyard current transformers	800/3000	26.7%	24 DEC 18

5.1.3 Mechanical Construction Work

The installation of mechanical equipment along with various piping systems on a large, complex construction project is extensive and requires a large number of highly skilled workers, which is the reason why this work represents a significant amount of the project's budget and time because it has to be done in a specified sequence. The push is to have a majority of the mechanical equipment installed at the time of electrical termination so that the motors and any instrument or control equipment (I&C) can be electrically connected after their respective cables have been pulled. The mechanical construction work on a large, complex construction project falls into five basic areas, as follows:

1. Underground Piping (including Firefighting Piping)
2. Mechanical Equipment
3. Process Piping
4. HVAC
5. Firefighting Piping

Underground Piping

On most large, complex construction projects, the tendency is to perform the excavation and installation of the underground piping after all of the foundations have been completed. If the diameter of the pipe is above 12 inches, the excavation work alone can involve limiting access to various areas of the construction site for a fairly long period of time, which can have a severe impact upon the project's schedule and budget. The EPC Construction Manager needs to work with both the Mechanical and Civil Engineers involved to mitigate this problem through the following steps:

Step No. 1 – Verify with both the EPC Mechanical and Civil Design groups what foundations will require internal piping to be installed, its exact location (coordinates), and where the interface points will be located on each of these foundations.

Step No. 2 – Establish a site drawing review by the site mechanical and Civil Engineers on the type of piping that will have to be installed and connected to each foundation. This should include establishing tie-in points to existing services, such as the heating steam, sewage, potable water, etc., and how they will be managed. After this review is completed, build the work scope— both civil and electrical—for each area where the underground piping will be installed and establish durations for all activities. Once completed, this work scope information should be compared to what was written in the project's CEP so that it can be updated, if required.

Step No. 3 – With the information provided from the Baseline Level 5 CPM schedule, now prioritize the piping systems to be installed and set up zones on the construction site for this work, which require equipment such as dump trucks, excavators or back hoes, compactors, etc. After the site is prioritized and laid out, establish a work area for the piping contractor to ensure that they have a place to uncrate the piping and keep a supply on hand so that the flow of work is maintained. This should include a number of welding bays with sufficient electric power so that small bore piping can be welded or assembled, depending upon the material, in parallel with the main work.

Step No. 4 – The EPC Civil Engineer will have to ensure that all of the excavation is performed according to the associated AFC drawings and the correct final elevation. After the area in the trench has been compacted to the specified density, another layer of sand or some other type of mixture will be placed prior to installing the pipe. The one critical item in this process is the verification of the material brought to the site for this layer of soil, which the EPC Construction Manager must not only have the approval of the EPC Civil Engineer but also that of the customer's Civil Engineer. After these approvals are received, then the installation of the layer can commence, and once completed, the area can be released for piping installation. The other critical item that must be checked and verified constantly during the installation of this underground piping is the slope of the pipe, as shown in Figure 5-8.

The progress of the piping, if it is HDPE (high-density polyethylene), PVC, or fiberglass, is usually measured by how many meters or feet are installed per day against the planned amount. If this is carbon steel piping, which are in sections that have to be welded, then the progress measurement is by "Diameter Inch," which is determined as shown in Table 5-5.

**SLOPE CALCULATION FOR A SECTION OF
UNDERGROUND PIPING**

(1) Slope Calculation: Rise/Run

Example:

			Length of		
Starting Elevation:	−72	Inches	Piping Run:	192 Inches	
Ending Elevation:	−68	Inches			
Rise:	4	Inches	Slope =	0.0208	Rise/Run
				2.08	Percent
				1.19	Degrees

Figure 5-8 Underground Piping Slope Calculation Example

Table 5-5 Underground Carbon Steel Progress Monitoring

WBS No.	Description	Pipe Diam. (Inch)	No. of Welds	Diameter Inch	Planned/Day Diam./Inch	Actual/Day Diam./Inch
3.1.1.1	Sect. A Potable Water	12	30	360	60	48
3.1.2.1	Sect. A Fire-Fighting Piping	20	25	500	100	80

The determination for the number of welds required is usually calculated by tabulating the number of welds specified in the respective piping isometric drawings for the particular piping system, which is performed by the EPC's Site Mechanical Engineer.

Step No. 5 – After a section is completed, the next step is to perform a leak test on that section, which is usually done with a hydro-test or a static leak test depending upon whether the pipe is under pressure or under atmospheric pressure during operation (drains). A hydro-test is a leak test procedure that involves filling a line with clean, treated water and then increasing the pressure of this liquid to a point that is above the system's operating pressure, as specified per the applicable engineering specifications. The static leak test, which is usually conducted on drain piping, involves filling the line with clean, treated water to a specified level on a vertical stand-pipe, which is monitored over 24 hours to determine if there has been a change that would indicate a leak. If a line is composed of carbon steel piping, then it normally has to be painted right after the testing is completed and before it can be covered with soil—called "backfilling."

Step No. 6 – The final step for underground piping is the backfilling of the trench with a specified soil mixture, which must be verified by both the EPC Civil Engineer and the customer's engineer. After a specific elevation, as confirmed by the surveyor, the soil mixture will be compacted to the specified density. In areas where a road or other type of crossing will be installed over the underground pipe, an additional form of support may have to be installed prior to final backfilling and compaction.

Mechanical Equipment

There are many systems present in a large, complex construction project, which lead to a large array of mechanical equipment that can be as simple as a valve or as complex as an HVAC chiller unit. The EPC Construction Manager has to align the Civil Work so that the required foundations and buildings are constructed in time to support the installation of the mechanical equipment as originally scheduled in the project's Level 5 Baseline construction schedule. The mechanical equipment installed depends solely upon the contract deliverables. For example, a large residential complex project that includes a recreation center and a solar power electrical system will have a much different type of mechanical equipment from a combined-cycle power plant project, which will require a large heat-recovery boiler, a gas turbine, a steam turbine, and all related auxiliary systems. However, all mechanical equipment has a standard method of installation, which is as follows:

1. **Erection and Preparation of the Foundation** – After the foundation has been poured and the concrete properly cured, the area where the equipment will be installed on the foundation is usually chipped down until the aggregate is showing, which is required for proper bonding at time of final grouting.
2. **Building Line Confirmation** – After preparation, the elevation, transverse, and axial alignment are verified to be as per the applicable AFC drawing and chalk lines established for the proper setting of the equipment.
3. **Lifting and Setting of the Equipment** – The large equipment should have a lifting plan that shows what lifting slings, shackles, etc., will be used to properly lift the equipment and set it in place on its foundation. After this is established, the large equipment can be brought to the site, usually along with a mobile crane of the right size (tons), and set in place on its foundation.
4. **Alignment of the Equipment** – During the installation of the equipment, it is usually adjusted so that it is installed on the correct building lines.
5. **Installation of Anchor Bolts** – After the alignment of the equipment is confirmed, the next step is to install anchor bolts at specified locations to lock the equipment in place.
6. **Grouting of the Equipment** – Once all of the anchor bolts have been installed, the final step involves installing shuttering around the equipment's foundation, which is usually a wooden-type box, for capturing the grout and then pouring the grout until the base of the foundation is adequately covered and filled with grout. If the grout is a cementitious type, it will have to be kept wet for a specific period of time until it has reached the desired strength, which is different than epoxy grout, which cures in a short time.

It is not important that the EPC Construction Manager know the exact technical details of each step, but he or she must fully understand the steps involved with installing mechanical equipment so that it can be managed properly. For example, if the subcontractor fails to chip the foundation, as described in the first step, the EPC Company will have to remove the equipment for this work, which is rework that will add an additional, nonrecoverable cost to the project's budget and possibly delay the project's planned completion date.

Process Piping

The next phase of the mechanical work is to connect piping, which is called process piping, from each piece of mechanical equipment to another piece. Once completed, this connection of piping comprises a specific system. Figure 5-9 is a simple illustration of a hot water heating system that shows how this work is accomplished.

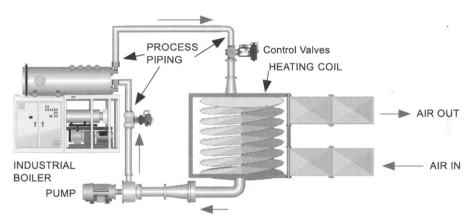

SIMPLIFIED HOT WATER HEATING SYSTEM FOR ILLUSTRATIVE PURPOSES

Figure 5-9 Illustration of a Simple Hot Water Heating System

The mechanical equipment of this simplified Hot Water System is as follows:

a. Industrial Boiler
b. Pump (Water)
c. Heating Coil
d. Control Valves

After the process piping is properly connected to each of these components, the final result is a hot water heating system. This methodology is standard when it comes to installation of process piping, but the EPC Construction Manager must ensure that the completion of each mechanical system is always in alignment with the project's Baseline Level 5 schedule. The progress of the process piping, which involves the welding of various pipe sections, must be closely monitored by the EPC Construction Manager and the EPC Site Mechanical Engineering Team. The two major standards of progress measurement are the following:

1. **Diameter-Inch** – This is similar to what is shown in Table 3-5, and the daily planned quantities must be monitored closely and action taken, if required, if the subcontractor starts to fall behind.
2. **Welds Accepted** – Each weld will require a specific type of Nondestructive Testing, which will be discussed in more detail in the next chapter of this book, before it can be considered acceptable. If it fails, the weld is considered rejected and must be redone, which is rework for the EPC Company. The industry standard is that this number of rejected welds on a system cannot exceed three percent. If this level is exceeded, the EPC mechanical engineer must immediately investigate, find the root cause, and take corrective action to bring this level back under the three percent level.

After the process piping is installed, the next step will be a pressure test at a specified pressure, which is in accordance with applicable engineering standards. This will ensure that the process piping will not leak during operation of the system. After this test is completed successfully, the process piping is painted, if required, or insulated with a specific, approved material.

Heating, Ventilation, and Air Conditioning (HVAC)

After a building is completed and released to the EPC Mechanical Engineering Team, the first system to be installed is usually the HVAC system because it involves a lot of ducting, as depicted in Figure 5-10.

This figure is a simplified illustration, but it clearly shows that the HVAC ducting has to be installed along with its supports prior to installation of the suspend ceiling for each room. After the chiller has been installed along with the necessary piping in the condensing unit, the installation of the ducting can start. The critical part of this installation is that the EPC's HVAC or mechanical engineer must ensure that the ducting is installed and aligned properly in each room and on each floor of the building. The EPC Construction Manager has to monitor the HVAC subcontractor's performance, which is normally done by the number of HVAC duct sections per day, planned versus actual amount

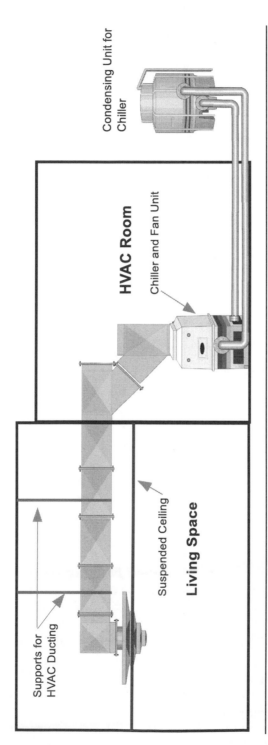

Figure 5-10 Simplified Illustration of a Simple HVAC Chilled Water System

installed.. If this is for a large building, which is common on large, complex construction projects, then the best way to establish zones with the most critical areas first according to the project's Baseline Level 5 schedule is as shown in Table 5-6.

Table 5-6 HVAC Ducting Installation Progress Monitoring

WBS No.	Location	Zone	No. of Duct Sections	Planned/Day Duct Sections	Actual/Day Duct/Sections	Total Installed to Date
3.4.1.2	1st Floor – Entry way	1	120	5	4	30
3.4.2.2	2nd Floor – Meeting rooms	2	140	7	5	40

After the HVAC ducting is installed, a duct leakage test is usually conducted to verify its integrity before it is covered with insulation. If there is a Chilled Water (CW) system, it will be installed and tested by the same method as discussed for process piping.

Firefighting Piping

In order to understand the various types of firefighting systems, it is essential that the EPC Construction Manager understand the basic rudiments of fire. In order for a fire to occur, it must have three elements, as follows:

1. **Ignition** – This is a form of energy that is capable of causing material to burst into flames, which is usually called "fire."
2. **Fuel** – This is any type of material that can burn, such as wooden chairs, clothes, electrical wiring coating, etc. Each type of material has a specific temperature at which it will burn and is a critical item in the design of a firefighting system.
3. **Oxygen** – If there is no oxygen, then there is no ignition because material will only burn in the presence of oxygen.

A firefighting system usually removes one of the three elements in order to control and extinguish a fire. For example, a fire in an electrical relay room is usually an electrical fire, which is best controlled and extinguished through the removal of oxygen. This is done through the use of carbon dioxide or another similar type of gas that suppresses the fire until the electrical energy is turned off. Since a person would suffocate in this environment, an alarm sounds just before this system is energized so that the space can be evacuated. This type of system is different from the standard water sprinkler firefighting system, which has nozzles that spray out a water type of fog when a fire breaks out in a room. This system works at reducing the ignition temperature of the items burning, which will put out the fire. The other major difference between these two systems is that the gas-suppression system is power by pressurized carbon dioxide (CO_2) cylinders in a special room through gas piping that is run into the room, but the water system is usually provided from a riser pipe that comes from an underground fire main, which is a pressurized water system.

The best way for the EPC Construction Manager to monitor progress is to break the building into zones, prioritize the rooms that require sprinklers or gas nozzles to be installed first, and then measure progress by the number of sprinklers or installed each week versus what was planned.

5.1.4 Instrument and Control (I&C) Construction Work

The last part of a large, complex construction project is the installation of the instruments required for monitoring and controlling the various systems that are completed as required for the project's deliverables. The first step is for the EPC Construction Manager to have the site EPC I&C Engineers review the Piping and Instrument Diagram (P&ID) for a system and establish the I&C work scope for it. This should continue for all systems so that the contractor chosen for this work knows the full extent of the I&C work that must be completed. An example of a simple water system P&ID is shown in Figure 5-11.

It is important that the EPC Construction Manager is familiar with the I&C instruments that must be installed, as follows (see Figure 5-11):

1. **TE – Temperature Element:** This is a temperature-sensing device that measures the temperature of a liquid or gas and then electronically provides that temperature to the operator in the control room or some other location, depending upon the project.

2. **FT – Flow Transmitter:** This is a flow-sensing device that measures the flow of a liquid or gas (gpm) and then electronically provides that flow information to a flow-controlling device (which is the case in this P&ID), and also to the operator in the control room or some other location depending upon the project.

3. **FC – Flow Controller:** This is a flow-controlling device that takes the input from the flow transmitter and matches it against an established set

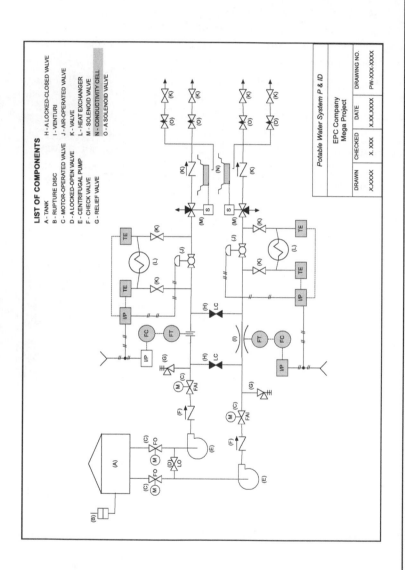

Figure 5-11 Simplified Potable Water System P&ID Illustration

point. If the flow is below the set point, it will send a signal to the valve to open more and increase the flow, which is the reverse if the flow is above the set point.

4. **I/P – Current to Pressure:** This instrument takes a current signal (milli-amps) and translates it into a pressure signal, which in this portable water system will cause the valve to open or close more to keep the flow at a specified amount.

5. **Salinity Cell:** This instrument measures the chloride content of the water, and if it is too high, then it will initiate the opening of a drain valve to dump the water.

In Figure 5-11, the items in yellow are all of the I&C instruments that must be installed before the system can be considered completed and ready for turn-over to the EPC commissioning team for operation. This installation also consists of connecting the required cabling to the instrument and then running that same wire into a panel located in the control room or the room that has the I&C control panel for the respective system being installed. A simplified example of the Potable Water System is shown in Figure 5-12.

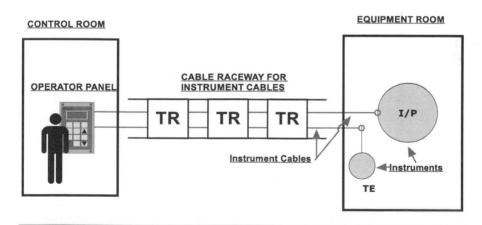

Figure 5-12   Simplified Illustration of Instrument Connection to Control Panel

By reviewing this simple illustration, the EPC Construction Manager can understand how to effectively monitor and control the progress of the I&C work onsite in the following manner:

1. Install all of the instruments required for the particular system and establish a required weekly amount as required to meet the completion date

specified in the project's Level 5 Baseline schedule. The Site EPC I&C Engineer will have to closely monitor this installation work so that it is completed as required.

2. If possible, during installation of the instruments, start installing the required cable racks for the specified electrical raceway. The same progress management tools discussed for the site Electrical Work should be applied, which is comparing the actual amount of cable trays in feet or meters per day versus the planned amount required to meet a specified completion date.

3. After the electrical raceway and instruments are installed, the installation of the instrument cable can start, and progress is measured as discussed for the site Electrical Work, which is comparing the actual amount of cable pulled in feet or meters per day versus the planned amount required to meet a specified completion date.

After all of the system instrumentation has been installed and connected to the control panel or an operator panel, the system is considered complete and ready for testing/operation by the EPC commissioning team. The EPC commissioning team takes over the system at this point from the construction team and will turn it over to the customer, once all of the required testing is completed.

5.2 Concluding Remarks

In this chapter, we examined all of the work that has to be performed by each engineering discipline and how to effectively manage it onsite. The EPC Construction Manager must also work diligently to coordinate the activities of all subcontractors onsite to ensure that there are no major disruptions in the performance of critical activities and those activities close to the project's critical path. In addition to this, the EPC Construction Manager has to ensure that the progress information is disseminated not only to the EPC Company's senior management in weekly and monthly reports, but also to the customer. Since most companies usually choose the less-expensive subcontractor, the EPC Construction Manager has to engage each subcontractor when they arrive onsite and carefully analyze their schedule to confirm that it is in alignment with the project's Level 5 Baseline schedule.

In the next chapter, we will see how the work onsite, which we discussed, will be brought into alignment with the Project's Quality Plan so that the deliverables produced will meet the quality standards set forth in the contract.

Chapter 6

Quality Management System

6.1 Quality Management System

The Quality Management System established onsite for the construction work is designed to implement the Projects Quality Plan (PQP) objectives through the use of various documents so that the deliverables produced meet the quality standards specified in the contract. In this chapter, we will examine these various documents along with their interrelationships and learn how to use the project's Work Breakdown Structure (WBS) to ensure that they remain in alignment with the Work Management System over the life of the project.

The Quality Management System is made up of a large number of documents, as depicted in Figure 6-1. The Quality Management System's final output is the ECC—**E**rection **C**ompletion **C**ertificate—for each of the systems being erected onsite, which will be discussed at the end of this chapter.

In order to understand how the Quality Management System is established onsite and effectively managed, we will start with "Approved for Construction (AFC) drawings" and then move onto each document in Figure 6-1.

6.1.1 Approved for Construction (AFC) Drawings

The EPC Engineering Department starts its design work for the project once the contract is signed, but each drawing produced must also be reviewed and accepted by the customer's engineers. This process, as shown in Figure 6-2,

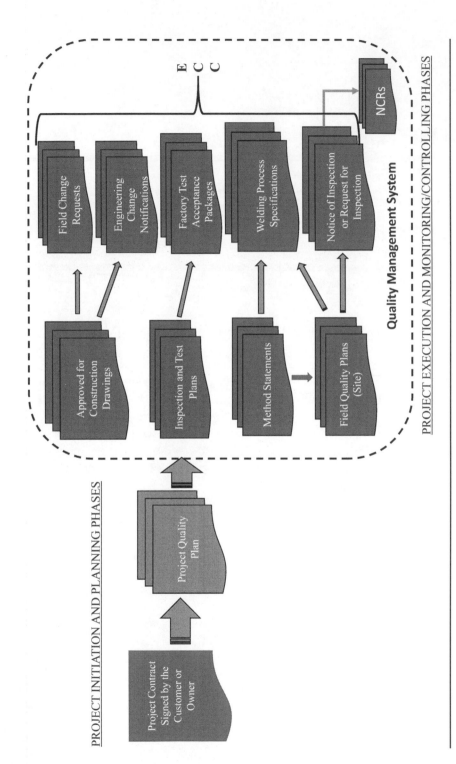

PROJECT INITIATION AND PLANNING PHASES

PROJECT EXECUTION AND MONITORING/CONTROLLING PHASES

Quality Management System

Project Contract Signed by the Customer or Owner

Project Quality Plan

Approved for Construction Drawings

Inspection and Test Plans

Method Statements

Field Quality Plans (Site)

Field Change Requests

Engineering Change Notifications

Factory Test Acceptance Packages

Welding Process Specifications

Notice of Inspection or Request for Inspection

NCRs

E
C
C

Figure 6-1 Quality Management System Documents

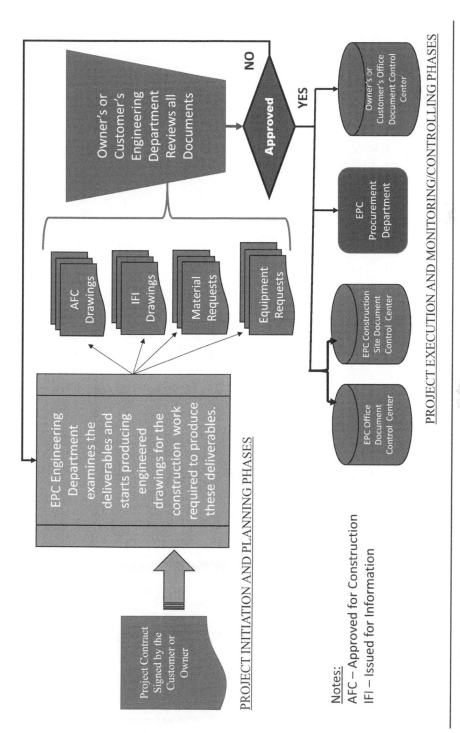

Figure 6-2 Engineered Construction Drawing Production Process

is lengthy for a large, complex construction project and must be carefully monitored by the EPC Program or Project Manager to ensure that the release of drawings is in accordance with the schedule requirements from the EPC Construction Manager and the EPC Procurement Manager.

If the deliverables for this project involve new technology, then the process shown in Figure 6-2 normally takes much longer, and the EPC Program or Project Manager has to work closely with the EPC Engineering Manager to prioritize the delivery of specific drawings and documents to the customer for immediate review and approval. On a typical large, complex construction project, just the drawings required for construction number in the thousands, which is why prioritization is critical for the approval process. This large amount of drawings also shows why the EPC Construction Manager needs to ensure that the Document Control Center is one of the first areas established onsite, along with a reliable form of document transmittal, so that all work is performed only with Approved for Construction (AFC) drawings.

6.1.2 Field Change Requests (FCRs)

On some projects, these documents are also called "Field Change Notifications (FCNs)" and are submitted by the EPC site quality team when something in the field has to be modified for proper installation. They must be approved quickly onsite so that work is not delayed. The critical item for this document is that the modification to the existing AFC drawing must not change the engineered design of the item or system being modified, which then allows the FCR or FCN to be reviewed by both the EPC and customer construction site engineers. The completed FCRs or FCNs must then be retained in the EPC's Quality Records for presentation to EPC Commissioning at the time of system or building turnover, once construction is completed.

6.1.3 Engineering Change Notifications (ECNs)

On some projects, these documents are also called "Engineering Change Requests (ECRs)" and are submitted by the EPC Site Engineering Manager when something in the field has to be modified, such as a piping system, building foundation, cable tray system, etc., which will change both the AFC drawing and the engineered design of this particular item. The process normally followed, which involves sending this document offsite, is shown in Figure 6-3.

Because the work involved with the ECN in Figure 6-3 cannot continue until it is approved, along with an approved solution, the EPC Construction

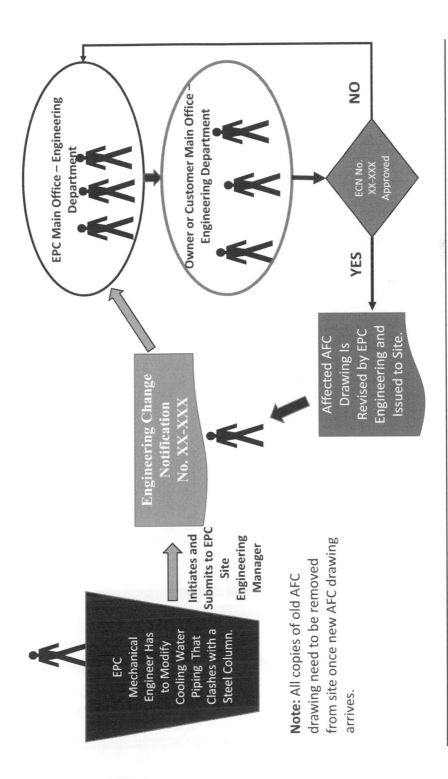

Figure 6-3 Engineering Change Notification Review and Approval

Manager has to consistently monitor its status and contact the EPC Site Project or Program Manager for assistance if it is not moving to a quick resolution. During this approval process, the EPC Construction Manager should be performing the following steps for this ECN, as follows:

Step No. 1 – Locate from the project's WBS structure the WBS element that is involved with this ECN and record it for project tracking and cost purposes

Step No. 2 – Talk to the subcontractor performing the piping installation and have its team provide a cost breakdown and schedule for the work described in this ECN

Step No. 3 – Contact the EPC Project or Program Manager and provide him or her with the information obtained in the first two steps

Step No. 4 – Update the project's Construction Execution Plan (CEP) to capture this ECN as a "lessons learned"

During this ECN approval process, the site Project or Program Manager on the project side should perform the following:

a. Take the information from the EPC Construction Manager and establish a ledger entry for this ECN in the project's Financial Ledger
b. File a change order internally, if this ECN was the result of an engineering oversight, to ensure that the costs will not go against the project's budget
c. Establish a weekly meeting with the EPC Site Engineering Manager and the Corporate Engineering Manager onsite to discuss engineering issues and include this ECN in that conversation

After this or other ECNs, the one critical item for the site QA/QC Manager is to ensure that the revised AFC drawing, which is usually issued by the EPC Engineering Department, is submitted to the subcontractor only after the old AFC drawing has been returned. This will ensure that no work is being performed onsite without the latest AFC drawing.

6.1.4 Inspection and Test Plans (ITPs)

These documents are usually developed by the EPC Engineering Department for all of the equipment that will be installed onsite during the construction process and are also submitted to the customer's engineers for their review and approval. This process, as shown in Figure 6-4, is actually an iterative process

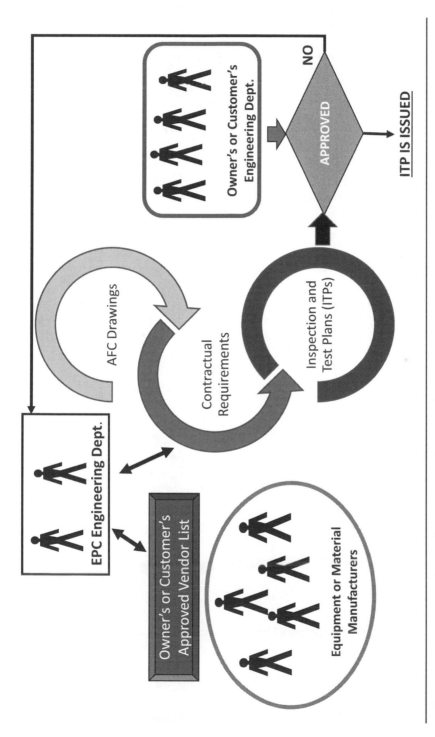

Figure 6-4 Inspection and Test Plan Review and Approval Process

that requires feedback from both the customer and the selected equipment or material manufacturers.

The process shown in Figure 6-4 is used to ensure that all equipment or material that will be a permanent part of the deliverables for this project will meet the customer's contractual requirements. For example, the ITP for a rotating pump will require that both the EPC and customer QA/QC inspectors witness specific points during the manufacturing and assembly of this pump in the factory. For example, the output of this pump at various pump speeds will not only be witnessed by both parties but also plotted so that the ITP for this activity can be signed off as complete. This will ensure that all applicable engineering specifications are being met so that the pump will perform as it should at the time it is operated.

6.1.5 Factory Test Acceptance Packages (FATs)

These are the factory tests for the equipment procured for the project, which are done in the factory and must be signed by both the EPC QA/QC inspector and customer QA/QC inspector prior to the equipment being shipped to the construction site. For example, the pump, which we discussed in the ITP section, would be operated in the factory and only released once the FAT is approved by the EPC and customer representatives. At the time of its shipment, a copy of the signed ITP and FAT will be included, along with all other documents required for custom's clearance at its port of entry. If the item is material, such as structural steel, this FAT will involve a review of the Nondestructive Testing (NDT) records for all of the fabrication welds along with a visual inspection of each weld to verify its integrity and compliance with applicable codes. After this is done, then another examination will be performed on the coating of each piece of steel as well as a dimensional check to ensure that it matches the specified fabrication drawing and EPC AFC drawing, which is critical before it is allowed to be shipped to the construction site.

6.1.6 Method Statements (MSs)

These documents, which are usually provided by the subcontractor performing the work, are submitted to the EPC Construction Manager and the EPC Site Engineering Team prior to the start of the work. The structure of an MS, which is for the water pump discussed in the ITP section of this chapter, is shown in Figure 6-5.

The Table of Contents shown in Figure 6-5 is the basic outline of a standard MS, which should contain the following sections:

XXX EPC COMPANY	RETIREMENT COMMUNITY BETA PROJECT			
	PARENT DOC. NO.	FSCP- XXX-QA-MS-NO2CWP-001		
	PROJECT DOC. NO	FSRP-YRG-CW0020-PMP-00240		
	WBS NO.	3.1.2.1	REV. NO.	1
METHOD STATEMENT FOR INSTALLATION OF POTABLE WATER PUMP			Page 1 of 6	

TABLE OF CONTENTS

Figure 6-5 Example of a Method Statement (MS) Table of Contents

1. **Objective** – This describes the purpose of the MS and the equipment or material covered, along with the work activity that will be performed, which is usually either installation or commissioning. The term "commissioning" refers to operating a piece of equipment for testing purposes only, which provides operation data that must be reviewed and accepted by both the EPC Company and the customer. In this example, it would be for the installation of the water pump, which would read as follows:

   The intent and purpose of this MS is for the installation of the water pump in accordance with applicable project specifications and the manufacturer's requirements.

2. **Scope of Work** – This section should provide a brief description of the work that the subcontractor will perform for the installation or operation of a piece of equipment or material, such as structural steel, conduit, cable trays, etc. In this example, its work scope for the installation of the water pump would read as follows:

   The work that will be performed for installing the water pump will include lifting the pump assembly, setting it on its foundation, aligning the pump assembly, and then aligning the motor to the pump after the base is grouted in position. This work will be conducted in accordance with all applicable AFC drawings and engineering specifications.

The EPC Construction Manager should ensure that the scope of work in the MS matches what is written in the contract.

3. **Resource Requirements** – This section should detail the resource requirements for the scope of work that the subcontractor will perform. The clearest way to present this information is in a tabular format, which is shown in Table 6-1 for our water pump example.

   This table should also provide the EPC's and customer's engineers with a quality checklist that can be used to confirm if the subcontractor is ready to perform the work.

Table 6-1 Table Showing the Resource Requirements for a Water Pump Installation

Tools, Equipment and Machinery	
Item	Description
1	Mobile Crane (15-ton capacity)
2	Rigging (10-ton capacity) per Lifting Diagram
3	Water and Pump on a Skid (supplied by Manufacturer)
4	Drill Motor for Drilling Holes in Concrete
Personnel	
1	Surveyor
2	Mechanics
3	Site Supervisor
4	Certified Crane Operator
5	Laborers
Material	
1	Cementitious Grout
2	Formwork
3	Anchor Bolts w/ Chemical Epoxy

4. **Lifting and Storage Requirements** – This section should specify where the material required for this work will be stored until the installation or erection work starts and provides the necessary vendor-approved lifting diagrams for the equipment or material to be installed. It should also clearly state that any material received onsite that does not meet the specified vendor specification or engineering requirements will be rejected and removed from the work area to preclude its usage onsite.

5. **Work Process** – This section of the method takes the work scope discussed above and breaks it down into steps that show the critical areas

of the installation process and the logical sequence that will be followed. The work process for the water pump, which is our example for this MS, would be as follows:

a. General

The installation of this water pump, which is on a skid, will be performed in accordance with the contract, under the supervision of a qualified EPC Engineer. This work will also be inspected at specific points by the customer as per the respective Field Quality Plan (FQP).

b. Location Verification

(1) The pedestal will be inspected first, and any areas requiring repairs will be completed by the EPC civil subcontractor.

(2) After the pedestal is accepted, a confirmation of the building centerlines and elevation will be confirmed by the surveyor. This will include clearly marking these positions on the pedestal's surface.

(3) The last step will be the verification and drilling of the anchor bolt holes required for proper anchoring of the skid after installation.

c. Pedestal Preparation

(1) The surface area of the pedestal, where the pump and motor skid will be set, will be chipped approximately .5 inch, to roughen and expose the aggregate for future grouting purposes.

d. Pump and Motor Skid Installation

(1) The pump and motor skid will be brought to the site and with the specified rigging, lifted and set on its foundation.

(2) After the pump and motor skid is set, the assembly will be aligned to the established building centerlines within specified tolerances.

(3) Once the alignment is completed, anchor bolts will be installed and epoxied in place as per the bolt manufacturer's requirements.

(4) After the epoxy for the anchor bolts has cured properly, formwork will be installed around the periphery of the skid and will be grouted in its final position with cementitious grout.

(5) The next step will be to align the pump to the motor in accordance with the manufacturer's specifications.

(6) The piping from this pump will now be installed, and the motor electrical wiring will be connected, as per the approved AFC drawings.

6. **Quality Records** – This section should specify the quality documents that will be produced during the installation of a piece of equipment or material on the construction site. The standard Quality Records will comprise the following documents:

a. Request for Inspection or Notice of Inspection

b. Inspection Checklist

These documents must be signed by both the EPC's inspector and the customer's inspector because they are considered legal project quality documents.

6.1.7 Welding Process Specifications (WPSs)

If there is any welding of piping or structural steel required during the installation of a piping system for a piece of equipment or the erection of structural steel, then there has to be a Welding Process Specification (WPS) for each type of welding. A generic example of this WPS is shown in Figure 6-6.

The WPS is usually written by a qualified welding engineer for the subcontractor, but the EPC Company also has to ensure that it is correct for the material being welded. The WPS is critical in ensuring not only that the welding being performed is in accordance with engineering specifications and codes but also that the welders are qualified to perform the welding, as specified by the WPS. The EPC QA/AC Manager should fulfill the following responsibilities when a subcontractor is chosen for the welding of either structure or piping:

a. The WPS, submitted by the subcontractor, is approved by both the EPC Company and the customer.
b. The WPS matches the material to be welded and what is specified by the submitted MS.
c. The Welder Qualification Test Record (WQTR) submitted by the subcontractor for each of the welders matches the WPS and has not expired.
d. The subcontractor has established an area onsite with the necessary weld coupons (pipe) or flat plate (structural) for recertification of each welder prior to the start of the welding work.
e. The completion of an audit of the welding machines brought to the site and confirmation that each machine has a valid certification that has not expired or will not expire during the progress of the welding work.
f. The verification with the subcontractor's welding engineers and the EPC Mechanical Engineers that a Weld Map from the AFC piping or structural drawings is available so that each weld is correctly identified for all applicable quality documents.
g. The confirmation that the subcontractor's welders completed their recertification by reviewing the NDTs of the weld performed by the welder, which are usually X-rays for piping. All the test results for each of the welders must be positive, and the EPC QA/QC Manager should only allow a welder about three times to pass. After that, he or she should tell the subcontractor that the welder has been rejected for welding onsite.

EXAMPLE OF A WELDING PROCESS SPECIFICATION (WPS) FORM

	Applicable Code for Welding (Check Box)
AWS D1.4/D1.4M:2011 (Structural Welding Code - Reinforced Steel)	
ASME B31.1-2016 (Power Piping)	

Welding Procedure Specification (WPS) No. _____

Contractor's Name _____
Authorized by _____   Revision No. _____

Additional PQR Identification
  (Work Specific) _____   Test Date _____

WELDING PROCESS: | FCAW-S | GMAW | SMAW | FCAW-G |
(Check One)

WELD TYPE: | Groove | Fillet |
(Check One)

JOINT TYPE: | T-Joint | Butt (Indirect) | Butt (Direct) |
(Check One)

Weld Position _____   Groove Type _____

Root Opening _____ Root Face _____   Groove Angle _____
Backing Ring : Yes _____ No _____   Backing Typ _____
Backgouging: Yes _____ No _____   Backgouging Method _____

TECHNIQUE: Stringers [   ]   Weave [   ]
(Check One)

BASE METAL:
  Material Specification _____   Grade _____
Welded to: Material Specification _____   Grade _____   Dissimilar [   ]
Maximum Carbon Equivalent _____ Bar size _____ Plate Thickne _____   (Yes or No)
  Steel Bar Coated?: Yes _____   No _____   Type of Coating _____
Pipe
  Maximum Chrome Content _____ Pipe Wall Thickness _____ Pipe Diameter or Schedule _____

PREHEAT/INTERPASS TEMPERATURES:
  Preheat/Interpass Temperature (Min) _____   Interpass Temperature (Max) _____

SELECTED FILLER METAL:
  AWS Specification _____   Low Hydrogen (Y or N) _____
  Filler Metal (If Not AWS) _____   Low Hydrogen (Y or N) _____

SHIELDING
  Gas: Single [   ]   Mixture [   ] Composition by % _____

ELECTRICAL INFORMATION:
  Current: AC [   ]   DCEP (+) [   ]   DCEN (-) [   ]
Mode of Transfer (GMAW): Globular [   ]   Spray [   ]   Short-Circuiting [   ]

WELDING PARAMETERS

Number of Pass	Electrode (Rod) Diameter	Type (AC or DC)	Range of Amps (Used)	Current Amt. of Volts	Electrical Stickout	Speed inch/min [mm/min.]	Joint Detail

Contractor (QA/QC Inspector) _____

Authorization Source _____   Date _____

Notes:
1. Weld Positions:
1G – Plate in flat position
2G - Plate in horizontal position
3G -Plate in vertical position
4G - Plate in overhead position
}   Structural and Tank Steel
1G Pipe - (rolled) in flat position
2G Pipe - (fixed) in horizontal position
5G Pipe -(fixed) in vertical position
6G Pipe - (fixed) inclined position at 45 degr
6GR Pipe - (fixed) inclined position with restriction ring
}   Piping
56

QA/QC Form W-01

Figure 6-6   Generic Example of a Welding Process Specification

h. The monitoring of each welder's performance by the number of welds rejected divided by the number of welds performed. For example, a welder is performing the welding of P-22 pipes and has completed 24 welds this week, but the NDT reveals that three of these welds have failed and will have to be rewelded once the defects are removed. This is a 12.5% rejection rate, which is not acceptable because 3% is the standard maximum rate allowed for welding in the field on the construction site. The subcontractor, after discussion with the EPC QA/QC Manager, can recertify this welder to see if he or she improves, but this type of situation cannot be allowed to continue for more than two weeks. The reason for this duration of two weeks is that this welder is currently producing a backlog of poor welds that have to be ground out to remove defects and then rewelded, which adds delay and cost to the project.

The WPSs, which usually comprise a large number on a complex, construction project, are the parent quality documents for all welding onsite, both piping and structural, and the EPC QA/QC must audit all welding subcontractors to ensure strict compliance at all times when the welding work is being performed. This audit must also include the subcontractor performing the preheating and post-weld treating of specific welds that require it because of the material being welded, which is also specified in the WPS.

6.1.8 Field Quality Plans (FQPs)

These documents are the backbone of the field Quality System because they specify not only which work is inspected only by the subcontractor and the EPC QA/QC engineers but also which work has to be inspected and approved by the customer's QA/QC engineers before work can proceed to the next step specified in the appropriate MS. A generic example of an FQP is shown in Figure 6-7, using the water pump that has been discussed previously.

In the example shown in Figure 6-7, just a few of the installation steps are shown so that the logical process, which is normally followed by the subcontractor and EPC Company, is clearly evident. The hold points, "H," are where the EPC Company has to formally invite the customer's QA/QC engineer for the final inspection—a Notice of Inspection or Request for Inspection (NOI or RFI), which will be discussed in the last section of this chapter. The EPC QA/QC Manager not only has to provide this invitation typically 24 hours in advance but also ensure that the subcontractor along with the EPC's engineer conducts an internal inspection so that the customer's QA/QC person will approve and sign off on the RFI when he or she inspects the completed work. If the inspection involves structural steel or pipe welding, a copy of the

CONTRACTOR:

EQUIPMENT IDENTIFICATION NUMBER: _____

PAGE ___2___ OF ___4___

CUSTOMER DOCUMENT NUMBER: FSCP-XXX-QA-MS-NO2CWP-001

EQUIPMENT LOCATION: Potable Water Service Building

APPLICABLE WBS No. _____

METHOD STATEMENT NO.

FSRP-YRG-CW0020-PMP-002240

Potable Water System Field Quality Plan (FQP) No. FSRP-FQP-PW0020-SYS-00240

MS Section	Step No.	Description of Activity	Date	PARTICIPANTS BY CATEGORY AND SIGNATURE				Remarks/Specific Documents/Engineering Specifications/Dwgs.
				Contractor	EPC Co.	Government (If Required)	Customer	
4	1	Lifting drawing and rigging verified to be correct for pump skid.		X	X		W	
5	1	Pump Skid is properly aligned to building centerlines.		X	X		H	AFC Drawing was checked and alignment matches specified dimensions.
5	2	Anchor bolts installed and exposied in place.		X	X		I	Curing time for the epoxy is three days.
5	3	Formwork installed and skid grounted.		X	X		W	Cementitious grout strength will be verified with cube samples taken at time of pour.
5	4	Pump to Motor Alignment Checked and Verified Correct.		X	X		H	Alignment is as per OEM Specifications.

Notes:
1. **W** - Stands for **Witness** - Customer does not need to inspect.
2. **H** - Stands for **Hold**, customer has to signoff before proceeding to the next step
3. **I** - Stands for **Information** only

Figure 6-7 Generic Example of a Field Quality Plan (FQP)

WPS and the welder's WQTR must also be attached to the NOI or RFI for the customer's engineer to review in advance of the inspection. The other item of importance is for the EPC Construction and QA/QC Manager to ensure that each FQP is assigned the correct WBS number for the project. This permits the QA/QC Manager to review the ongoing six-week Level 3 lookahead schedule and compile the number of inspections that must be completed in the next six weeks to keep the work moving forward. This compilation must be provided in a monthly report that the EPC Construction Manager can review to determine if any of the subcontractors are not completing their work on time, which is quickly determined by reviewing the planned inspection date versus the actual inspection date in the monthly QA/QC internal report. On a complex construction project there are typically hundreds of FQPs for all of the work to be performed, and without alignment to the project's baseline schedule, a correct site delivery date for each WPS to support the work can either be overlooked or forgotten. The EPC QA/QC Manager has to always remember that a WPS submitted by a subcontractor has to be reviewed and approved by the engineers from both the EPC Company and the customer. This process can take up to two months unless it is tracked and pushed by the EPC QA/QC Department with assistance from the Engineering Department, which adds greater priority to ensuring that the correct WBS number is on each FQP before it is submitted for approval.

6.1.9 Notice of Inspection or Request for Inspection (NOI or RFI)

As discussed in the previous section, the Notice of Inspection or Request for Inspection (NOI or RFI) is established by the inspection requirements of the corresponding FQP. This process is shown in Figure 6-8, which clearly shows the two distinct areas of inspection: the "Witness or Information Point" (where the EPC subcontractor has to prepare for the customer's inspection) and the "Hold" (where the customer, along with the EPC inspector and their subcontractor, performs the final inspection).

An example of a typical RFI is provided in Figure 6-9, which clearly shows a signature for all the parties involved with the inspection, as per the process shown in Figure 6-8, in addition to a description of the work and inspection.

The document and drawing information is critical and must be correct for the work being performed. For example, if this RFI was an inspection of the anchor bolt locations for a steel column, and the project's Master Drawing List (MDL) shows the AFC drawing having a Revision 3, then the RFI must also show the same information for this inspection. It is the responsibility of the site

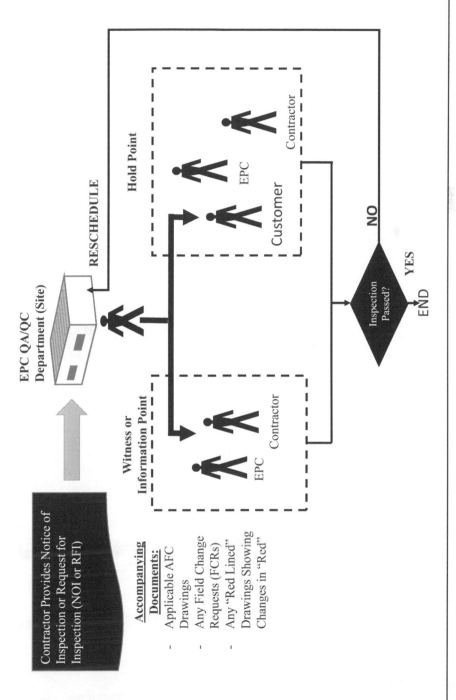

Figure 6-8 Inspection of Work Process During Construction

Request for Inspection (RFI)						
EPC CONTRACTOR NAME: _____ **RFI NO:** _____						

EPC Site Management Discipline*	Civil	Mech.	Elect.	I&C	QA/QC	Com.
Customer Site Discipline*	Civil	Mech.	Elect.	I&C	QA/QC	Com.

* - Circle the Appropiate Discipline Involved with Inspection

Field Quality Plan No.	
Reference Documents	AFC Drawing Nos:
	Documents:
Description of Work	
Inspection Scope	

Date of Inspection	
Site Location	

Acknowledgment of RFI Receipt

	Contractor	EPC	Customer
Name			
Signature			
Date			

Results of Inspection

Rejected		Accepted W/O Comments		Accepted W/ Comments	

Comments:

Completion of the RFI

	Contractor	EPC	Customer
Name			
Signature			
Date			

Figure 6-9   Example of a Typical Request for Inspection

EPC QA/QC Department and the EPC inspecting engineers to confirm that the necessary documents, such as drawings, procedures, etc., have been submitted to the customer's inspectors prior to the inspection.

6.1.10 Nonconformance Report (NCR)

This document is only produced by the site EPC QA/QC Manager if there is failure by the subcontractor to follow a specific quality document prior to the commencement of the work, which leads to major rework or replacement of specific items for correction of this mistake. For example, in the anchor bolt inspection discussed in the previous paragraph, if the subcontractor used an AFC drawing that had Revision 1 instead of Revision 3, the inspection will reveal that the anchor bolt locations are now in the wrong location with respect to building centerlines. The customer and the EPC are now faced with a major delay because the current anchor bolt location will require EPC Engineering in the corporate office to evaluate the current situation and determine if it is acceptable or has to be fixed. The EPC QA/QC Manager issues an NCR against the subcontractor for this quality violation, and it can only be closed once the engineering decision has been implemented on site, which also must be approved by the customer's Engineering Department. It is also the responsibility of the EPC QA/QC Manager to notify the EPC Construction Manager and the Program or Project Manager of any NCR issuance that is due to its potential cost impact, schedule impact or both. The quality nonconformance that was just discussed could possibly lead to a delay in construction for a building that is on the project's critical path, which could have a large schedule impact on the project's completion date, along with its planned budget.

6.2 Erection Completion Certificate (ECC)

The main objective of the Quality System onsite is to ensure that the quality of each deliverable is as per the contractual and customer requirements at the time of completion, which is what the Erection Completion Certificate (ECC) confirms. This process starts a few months prior to the completion of erection or construction of a system or structure because all associated quality documents must be reviewed for completion and compiled into one package, which accompanies the final NOI or RFI that is presented to the customer for the final walkdown. This process is shown in Figure 6-10.

The final walkdown of a system, as shown in Figure 6-10, involves all three parties—the customer, the EPC Company, and the subcontractor. The intent of

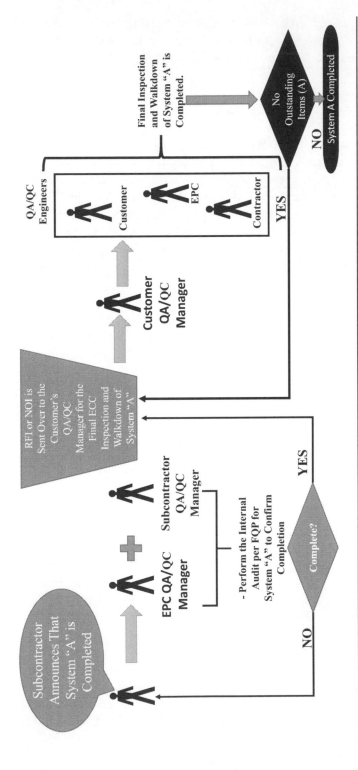

Figure 6-10 The Erection Completion Certification Process for a Construction Site

this final inspection and walkdown is to generate a punch list, if there are any findings, which is then segregated into three categories, as follows:

1. **Category A** – This finding must be fixed before the system can be operated and results in the ECC remaining open until all of the "A" findings have been closed by the subcontractor, which the EPC QA/QC Department must confirm before it issues another RFI or NOI to the customer's QA/QC Manager for the 2nd final inspection. If the final walkdown finds one or more Category A findings, the customer rejects the RFI or NOI and returns it to the EPC QA/QC Manager for resubmittal after the subcontractor fixes all of these items.
2. **Category B** – This finding is an item that must be fixed before the system can be operated, and is minor in nature. For example, if an inspection of the potable water system revealed that a valve wheel was missing on a valve, this item is required for operating the system, but it does not prevent closure of the ECC for the potable water system. This item would be categorized as Category B on the final walkdown punchlist.
3. **Category C** – This finding is an item than be fixed after the ECC has been closed, but it must be completed before the project is closed and the system is handed over for operation by the customer. For example, that potable water system, which was used in previous examples, may require some additional insulation and painting in specific areas of a building prior to the customer accepting the building for occupation. The punch list will put this finding down as Category C.

The EPC QA/QC Manager must always inform the EPC Construction Manager when an ECC fails the final walkdown inspection because the items to be fixed are rework by the subcontractor, which can add additional cost to the project's budget and potentially push out the final acceptance date for the project's critical path deliverables. The EPC Commissioning Team members are the ones who will be involved with receiving the systems from construction as they are completed and once the respective ECC is signed off on without any outstanding "A" items. After all of the systems have been operated and tested successfully by the EPC Commissioning Team, the next phase is the turnover of all project deliverables to the customer. On a complex construction project, the buildings or structures usually do not have a formal ECC, but any of the building support systems, such as potable water, firefighting, lighting, HVAC, etc., will have a formal ECC that must be signed off on and closed.

The Quality System implementation and management requires continuous daily engagement by the EPC QA/QC Manager and the onsite QA/QC team because they need to be monitoring the subcontractor's work along with what has been done so that any quality defects are identified and corrected early.

6.3 Concluding Remarks

In the next two chapters, we will see how the Project Management System, the Work Management System, and the Quality System, when linked through the project's WBS, can be used to effectively monitor, control, and manage a large, complex construction management project from the first scoop of dirt to the last weld on a section of pipe.

Chapter 7

Bringing It All Together*

7.1 Retirement Community Beta Project

7.1.1 Project Initiation

The Project Manager, Bob, for the Comet EPC Company received a phone call from the company's Senior Vice President of Sales and Marketing late on Friday regarding a unique opportunity for the company to get into the "green" residential community market, which was one of their Chief Executive Officer's (CEO's) visionary goals for the company that was discussed at the company's last stockholders meeting.

Pre-Bid Analysis

The opportunity arrived in the Project Manager's "in-box" electronically just as the Engineering Manager walked into his office and mentioned that they both should look at this project picture, which the Senior Vice President of Sales sent over. The customer—50+ for All Seasons—a new one for the company, had a reputation in the "green" residential community market as one of the best builders and was looking for a reputable EPC Company to build a new "green" community for them. They both were surprised when they looked at the graphical

---

* All of the names of people, places, companies, and any related facts presented in this chapter are fictitious.

layout of the project, which was only for information but which revealed the complexity of the scope that the EPC Company would have to manage if they wanted to be successful. This layout, as shown in Figure 7-1, contained not only two apartment-type buildings but also a medical clinic and a sports complex, which was going to be powered primarily using wind and solar energy.

The complexity was not something that the Project and Engineering Managers felt could not be managed, provided that the project was properly aligned with a strong degree of governance at each level, but it would involve

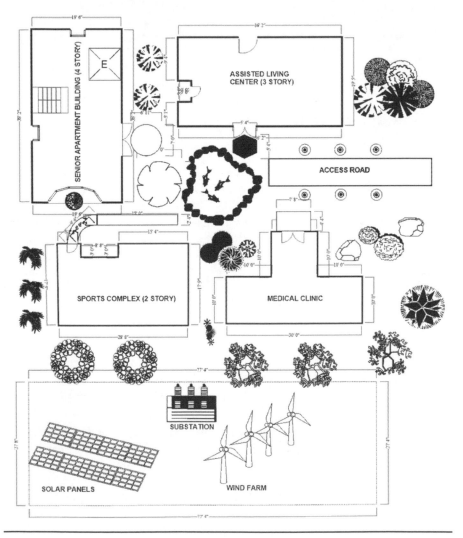

Figure 7-1   Graphical Layout of the Retirement Community Beta "Green" Project

breaking this project up into a group of "mini-projects." The Project Manager had previous experience working on large, complex construction projects for another company, which he explained to the Engineering Manager. They called the Senior Vice President of Sales and Marketing and told him that they felt that the company should move forward with this opportunity.

The next morning, the Senior Vice President of Sales and Marketing informed the Project Manager that she was able to get the company put on the bidder's list for this new Retirement Community Beta "Green" Project, and that the pre-bid meeting was scheduled for October 14th. This gave the Project Manager just three weeks to build an organizational chart for the group he needed complete the bid and a presentation for senior management, which was required before he could receive an internal "Authorization to Proceed" from the company's senior management. This internal "Authorization to Proceed" was required because one bid for a large, complex construction project usually requires a substantial amount of internal resources and operating capital, which has to be approved by the company's Executive Committee before anyone can attend a pre-bid meeting. After a meeting with each of the department heads to discuss the resources required to develop a bid for this project, the Project Manager, Bob, was able to develop the organizational chart, as shown in Figure 7-2.

An examination of this organizational chart also reveals resources, listed under each department head, from that department that will be required for the compilation of this bid, once it starts, which will also be communicated to the EPC Company's senior management in the presentation for this project. After this is completed, the next step requires establishing a Work Breakdown Structure (WBS) for the project bid compilation process, which also starts the project initiation phase for this project. This is so that the EPC Company's senior management can have an overall picture of how this process, which begins with receipt of a project bid and ends with a bid submission to the customer for the Retirement Community Beta "Green" Project, is structured. This WBS is shown in Figure 7-3.

Because the WBS shows a clear decomposition of the work to be performed, along with a logical process of completion, it becomes the basic building block for the Project Management System discussed in Chapters 2 to 4 and permits the compilation of a schedule. This schedule is compiled by taking each activity in the WBS and adding a duration and cost to it until all of the activities have been completed. The Project manager, Bob, does exactly this with the WBS shown in Figure 7-3 and produces the schedule for this project's bid compilation, which is shown in Figure 7-4.

This schedule is a called a "PERT" because it was derived using the Program Evaluation Review Technique (PERT), which is a process where one starts with

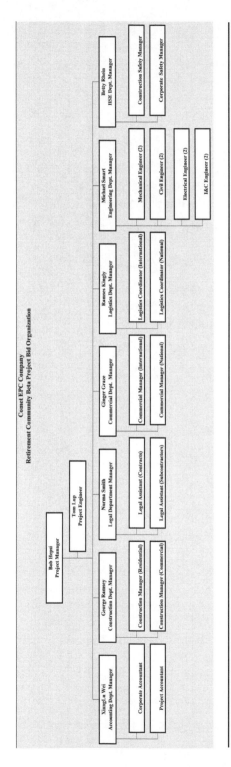

Figure 7-2 Retirement Community Beta Retirement "Green" Project Bid Organization

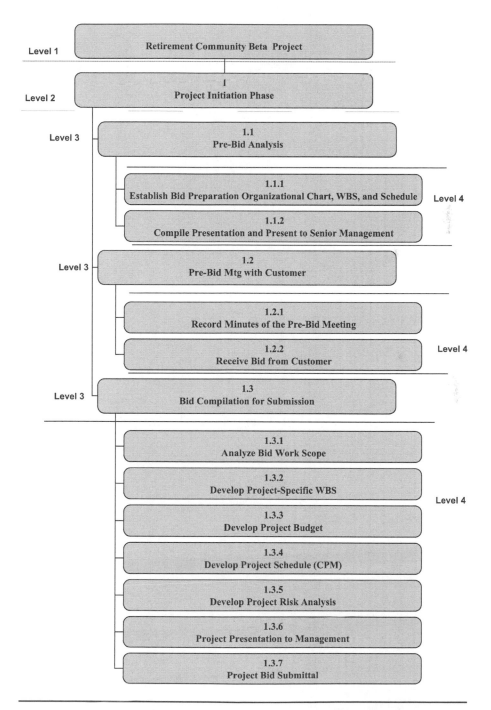

Figure 7-3 Retirement Community Beta "Green" Project Bid Compilation Work Breakdown Structure (WBS)

WBS	Task	Assigned To	Start	End	Duration	2017 Sep	Oct	Nov	Dec
	Retirement Community Beta Project	Project Manager	9/14/17	12/22/17	83				
1	Project Initiation Phase		9/14/17	12/22/17	83				
1.1	Task		9/14/17	10/14/17	25.5				
1.1.1	Establish Bid Preparation Organizational Chart, WBS, and Schedule	Project Manager	9/14/17	10/2/17	16				
1.1.2	Task	Project Manager	10/3/17	10/12/17	8				
1.1.3	Receive Senior Management Approval	Project Manager	10/13/17	10/14/17	1.5				
1.2	Attend Pre-Bid Meeting w/ Customer	Commercial Manager Project Manager Senior VP Sales and Marketing	10/14/17	10/16/17	1.5				
1.2.1	Record Minutes of the Pre-Bid Meeting	Commercial Manager	10/14/17	10/14/17	0.5				
1.2.2	Task	Senior VP Sales and Marketing	10/15/17	10/16/17	1				
1.3	Bid Compilation for Submission	Program Manager and Project Manager	10/17/17	12/22/17	56				
1.3.1	Analyze Bid Work Scope	Bid Compilation Team	10/17/17	10/30/17	12				
1.3.2	Develop Project-Specific WBS	Bid Compilation Team	10/31/17	11/21/17	18				
1.3.3	Develop Project Budget	Bid Compilation Team	11/22/17	12/9/17	15				
1.3.4	Develop Project Schedule (CPM)	Bid Compilation Team	11/22/17	12/9/17	15				
1.3.5	Develop Project Risk Analysis	Bid Compilation Team	10/31/17	12/13/17	36				
1.3.6	Project Presentation to Senior Management	Bid Compilation Team	12/14/17	12/19/17	5				
1.3.7	Project Bid Submittal	Bid Compilation Team	12/20/17	12/22/17	3				

Figure 7-4 Retirement Community Beta "Green" Project Bid Compilation Schedule

a fixed end date for a milestone, which for this project, at this time, is the pre-bid meeting date of October 15th, and then works backwards to develop the schedule. One critical item that the Project Manager utilized in this process was determining the relationships among the activities, which are normally defined in scheduling, as shown in Table 7-1.

Table 7-1 Scheduling Relationship Table

Relationship	Abbreviation	Explanation
Finish-to-Start	FS	The first activity must be finished before its succeeding activity can start.
Start-to-Start	SS	The first activity and its succeeding activity can start at the same time.
Finish-to-Finish	FF	The first activity and its succeeding activity must finish at the same time.
Start-to Finish	SF	The first activity cannot start until its succeeding activity has been completed.

For example, in the schedule shown in Figure 7-4, Activity Nos. 132 and 135 have a Start-to-Start (SS) relationship because they are independent of each other and can start immediately after the project's bid workscope has been analyzed. The other important item in a schedule is to clearly establish the person or persons within the EPC Company who have responsible for each activity (see Figure 7-4) because it is critical for effective communication within the project's organizational structure.

Once the Project Manager had time and duration, along with resources, for each activity, the next step was working with the Accounting Department to put together a preliminary budget for the pre-bid compilation work, which senior management usually requested at these "Pre-Bid" meetings because these funds came out of the company's internal operations budget.

On October 13, 2017, the Project Manager, Bob, presented the Retirement Community Beta Project to the senior management of the EPC Company and highlighted how it fit into the company's goal of moving into the "green" residential community construction market. After the presentation, the Senior Vice President of Construction stated that the EPC Company's senior management was excited about this project and gave their "Authorization to Proceed" with bidding for the construction of this Retirement Community. The Senior Vice President of Construction also told Bob that he is the Sponsor for this important project and to contact him directly if Bob encountered any internal or external problems that would jeopardize the success of this project.

Attend Pre-Bid Meeting with Customer

The pre-bid meeting was conducted on October 14th by the 50+ for All Seasons Project Management Office (PMO) at a conference center in a city that was about two hours away from the area where the Retirement Community would be constructed. During this meeting, a timetable was presented by the customer's Program Manager (see Figure 7-5).

RETIREMENT COMMUNITY BETA PROJECT BID

TIMELINE FOR ALL QUALIFIED BIDDERS TO FOLLOW

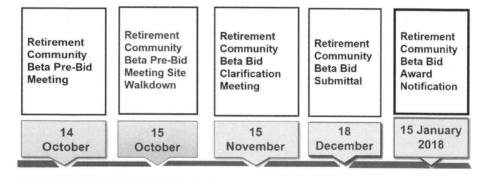

Retirement Community Beta Pre-Bid Meeting	Retirement Community Beta Pre-Bid Meeting Site Walkdown	Retirement Community Beta Bid Clarification Meeting	Retirement Community Beta Bid Submittal	Retirement Community Beta Bid Award Notification
14 October	15 October	15 November	18 December	15 January 2018

Figure 7-5   Retirement Community Beta Timetable

The one item that came up in the meeting regarded the interfacing with other companies—that is, for potable water tie-in, sewage tie-in, access roads, and, especially, the renewable energy park for this complex. The Program Manager stated that they will have one Site Project Manager, who will be responsible for coordinating all of these interfaces and effectively managing each of them so that there will be no impact on the project or on the EPC Contractor. The customer's Program Manager also stated that there will be questions from each of the contractors after they review the bid package; therefore, 50+ for All Seasons has scheduled a Bid Clarification Meeting at their headquarters on November 15, 2017, for all contractors, and anyone who does not attend will be excluded from the bidding. The last item was that the date of December 18th for bid submittal was fixed, and any bids submitted after this date would not be accepted by 50+ for All Seasons. Once the presentation was completed, the rest of the meeting was spent with the Program Manager and other members of the PMO answering questions from each of the contractors present.

The meeting was conducted over a span of two days, with the second day being a walkdown of the future construction site for all contractor representatives, which the Project Manager attended. During this walkdown, the customer's Construction Manager first gave everyone present a site layout drawing to be presented to the local government office by 50+ for All Seasons for the Site Work Permit, which they planned to receive by February 1, 2018. As the group followed the site layout drawing under the direction of the customer's Construction Manager, the Project Manager, Bob, began to see that this was going to be a much more complex project than he originally thought and had experienced in previous projects, which was substantiated by the pictures he was permitted by the customer to take during the walkdown.

Bid Compilation for Submission

After the Senior Vice President of Sales and Marketing and the Project Manager, Bob, returned with other members of the Comet bid team, a meeting was called for all department heads to discuss the newly received retirement community bid package and to establish assignments for each department so that the EPC Company could ensure that they would be ready for bid submission on December 18, 2017. Because this was the start of the Project Management System, the Project Manager, Bob, decided to use its inputs to develop a checklist for the bid compilation, which is shown in Table 7-2.

The start and end dates for Table 7-2 are derived from the schedule established for the bid compilation work, which is shown in Figure 7-6.

The milestone for this schedule is Activity No. 1.5 Retirement Community Beta Bid Submittal, which has to be completed on December 18, 2017, because the customer made it very clear that no late submissions would be accepted. Table 7-2 could also be used by the Project Manager, Bob, to update progress on a daily basis so that each morning he can update all members of the Bid Compilation team on the current performance of the team and discuss any obstacles or successes that they each member is experiencing.

One of the obstacles encountered during an examination of the bid from the customer was the requirement to procure, install, and commission a small solar-generating power plant, which was something new to the Comet EPC Company. However, the company had recently hired a Program Manager, Sandy Xandau, who had a lot of experience managing the installation of renewable energy power plants in the utility business, and she was assigned to work with the Project Manager on all projects that had some type of renewable energy deliverables. Sandy also informed Bob that if they were successful in being awarded this project, she would show him how to break the project into a

Table 7-2 Bid Compilation Responsibility and Progress Checklist

Project Management System Input	Department or Departments Responsible	Start Date	End Date	Current Progress (%)	Comments
Project Contract (Scope)	• Engineering Dept. Manager • Commercial Dept. Manager • Legal Dept. Manager • Construction Dept. Manager • Project Manager	10/17/17	10/30/17		
Project-Specific WBS	• Project Manager • Engineering Dept. Manager • Construction Dept. Manager • Project Engineer • Commercial Dept. Manager	10/31/17	11/21/17		
Project Budget	• Accounting Dept. Manager • Construction Dept. Manager • Engineering Dept. Manager	11/22/17	12/9/17		
Project Stakeholder List	• Project Manager • Commercial Dept. Manager • Project Engineer • Legal Dept. Manager	12/01/17	12/07/17		
Project Schedule (Level 4 or 5)	• Project Manager • Engineering Dept. Manager • Construction Dept. Manager • Project Engineer • Commercial Dept. Manager	11/22/17	12/9/17		
Project Resource Requirements	• Project Manager • Engineering Dept. Manager • Construction Dept. Manager • Project Engineer • Commercial Dept. Manager	11/22/17	11/28/17		
Project Engineering Requirements	• Project Manager • Engineering Dept. Manager • Construction Dept. Manager	11/22/17	11/28/17		
Project Risk Register	• Project Manager • Engineering Dept. Manager • Construction Dept. Manager • Project Engineer • Commercial Dept. Manager	10/31/17	12/08/17		
Project HSE Requirements	• HSE Dept. Manager • Commercial Dept. Manager • Project Manager • Project Engineer	12/10/17	12/16/17		

group of projects so that the governance and alignment with corporate goals for these types of projects could be established at every level.

To accelerate the completion of this bid, all members of the Bid Compilation team agreed to follow the Program Manager's advice, which was to break up

#	Task	Assigned To	Start	End	Duration	%	2017 Sep Oct Nov Dec
	Retirement Community Beta Project	Project Manager	9/14/17	12/17/17	78.5		
1	Project Initiation Phase		9/14/17	12/17/17	78.5		
1.1	Pre-Bid Analysis		9/14/17	10/13/17	25		
	Attend Pre-Bid Meeting w/ Customer	Commercial Manager, Project Manager, Senior VP Sales, and Marketing	10/14/17	10/15/17	1.5		
1.3	Bid Compilation for Submission	Project Manager	10/15/17	12/16/17	51		
1.3.1	Bid Work Scope Analysis	Bid Compilation Team	10/15/17	10/29/17	12		
1.3.2	Project-Specific WBS	Bid Compilation Team	10/29/17	11/21/17	18		
1.3.3	Project Budget	Bid Compilation Team	11/21/17	12/12/17	17		
1.3.3.1	Project Resource Plan	Bid Compilation Team	11/21/17	11/28/17	5		
1.3.3.2	Construction Execution Plan	Bid Compilation Team	11/30/17	12/12/17	10		
1.3.3.3	Project Procurement Plan	Bid Compilation Team	11/23/17	12/3/17	8		
1.3.4	Project Schedule (CPM)	Bid Compilation Team	11/21/17	12/9/17	15		
1.3.5	Project Risk Analysis and Plan	Bid Compilation Team	10/29/17	12/8/17	32		
1.3.6	Project Organizational Chart	Project Manager	11/21/17	11/30/17	7		
1.3.7	Project Communication Plan	Bid Compilation Team	11/30/17	12/7/17	6		
1.3.7.1	Project Stakeholder List	Bid Compilation Team	11/30/17	12/7/17	6		
1.3.8	Project HSE Plan	HSE Dept. Manager	12/9/17	12/16/17	6		
1.3.9	Project Quality Plan	QA/QC Dept. Manager	10/29/17	11/5/17	6		
1.3.10	Project Presentation to Management	Bid Compilation Team	12/8/17	12/14/17	5		
1.4	Project Bid Submittal (Management)	Bid Compilation Team	12/14/17	12/16/17	2		
	Retirement Community Beta Project Bid Submittal	Commercial Manager, Project Manager, Senior VP Sales and Marketing	12/16/17	12/17/17	1		

Figure 7-6   Bid Submittal Schedule for Retirement Community Beta

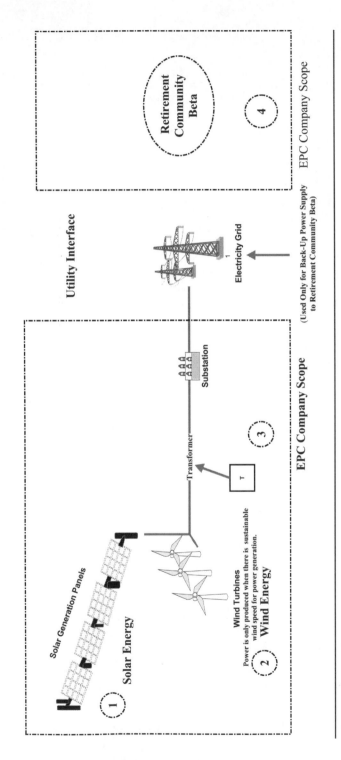

Figure 7-7 Breakdown of the Retirement Community Beta into Four Projects

the Retirement Community workscope into four projects because each had a unique set of deliverables. This breakdown, as shown in Figure 7-7, permitted the team to also establish those areas where an outside source of information would be required in the bidding process.

The compilation of the workscope for this particular complex construction project was made much easier by examining each of the four projects and identifying the common areas of scope versus those that are unique to the particular project, which is shown in Table 7-3.

An examination of this table by the Bid Compilation team revealed that the Comet EPC Company had sufficient experience in the area of building construction and electric system distribution work from previous projects, which was captured in the company's Construction Execution Plan (CEP), to develop most of the bid. However, it had no experience in constructing the renewable-energy electric-generating facilities, which were required for bidding the work required in Project Nos. 1 and 2.

The Project Manager, Bob, remembered that the company had just hired a new Project Manager, Nancy Anjou, solely for renewable-energy projects, who had past experience managing renewable-energy construction projects for a utility, and he was sure that she could manage the compilation of the bid for the work required for the Project Nos. 1 and 2 deliverables.

After the new Project Manager, Nancy, reviewed the renewable energy area of the Project Contract workscope, she informed Bob that the Comet EPC Company would have to contract this work out, but she knew of at least three companies that had this type of construction experience and would be willing to provide input for bidding the work. After Bob contacted and received permission from the Senior Vice President of Construction to proceed, Nancy started compiling the required bid information for Project Nos. 1 and 2, with the assistance of a small project team quickly assembled for this purpose.

After the Bid Compilation team completed the work breakdown and each of the four projects were assigned a specific Project Manager to coordinate the compilation process, the WBS was established, as shown in Figure 7-8.

In the WBS shown in Figure 7-8, the notation "Contractor," in parentheses, has been added to both Project Nos. 1 and 2 so that the Project Manager, Bob, along with the rest of the Bid Compilation team, realized that part of the bid estimates for the deliverables of these two projects would have to be provided by a subcontractor. The reason for this notification is that requiring bid information from a subcontractor can possibly delay a bid submittal and must be carefully managed for the following reasons:

1. Support from the Commercial Department is usually required to qualify a subcontractor if they are not on the approved vendor list, which would

Table 7-3 Activity Breakdown by Project for the Retirement Community Beta

Project No.	Site Location	Common Construction Activities	Unique Construction Activities Related to the Project
1	Solar Energy Electrical-Generating Area	• Civil Site Preparation • Civil Excavation Work • Civil Foundation Erection • Electrical Earthing • Firefighting System Installation • Site Drainage System Installation	• Panel Support Structure Erection • Solar Panel Installation • Electric Power Cable Installation • Electrical Duct Bank or Trench Construction • Final Site Preparation After Panel Installation • Commissioning and Testing
2	Wind Energy Electrical-Generating Area	• Site Access Roads • Permanent Site Access Roads • Site Perimeter Fence	• Structure and Platform Erection • Equipment Installation • Electric Power Cable Installation • Electrical Duct Bank or Trench Construction • Final Site Preparation After Wind Generator Installation • Commissioning and Testing
3	Electric Distribution and Substation		• Building Erection • Electrical Equipment Installation • Building Services (including HVAC) • Electric Power Cable Installation • Electrical Duct Bank or Trench Construction • Final Site Preparation After Panel Installation • Commissioning and Testing
4	Retirement Community Beta		• Multi-Story Residential Building Construction (2) • Medical Building Construction • Medical Equipment Installation • Sports Complex Building Construction • Swimming Pool Construction Inside Sports Complex • Walking Track Construction Inside Sports Complex • Building Services installation for All Buildings • Electrical Distribution Construction for the Retirement Community Beta Complex • Commissioning and Testing of All Building Systems

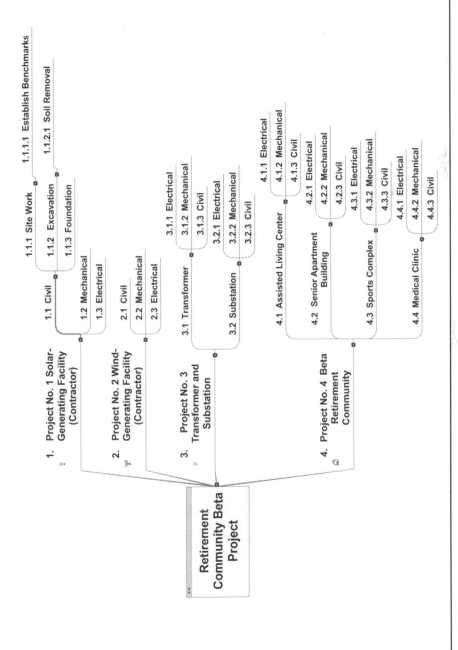

Figure 7-8 Retirement Community Beta Project Work Breakdown Structure (WBS)

COMET EPC COMPANY INC.

BID TABULATION FORM

Project Phase:	Initiation	Project No.: 4	
Project Name:	Retirement Community Beta	**Project Description:**	Installation of One
Location:	Tempe, Arizona		Bank of Solar Panels
Author/Group:	Renewable Energy		to Generate Electricity
Bid Due Date:			

BID QUANTITIES

Bid Item No.	WBS No.	Item Description	Qty.	UNIT	EPC's Estimate UNIT COST/HOUR	EPC's Estimate Number of Days	EPC's Estimate BID	CONTRACTOR A (Solar Generation Facility) UNIT AMOUNT	CONTRACTOR A BID	CONTRACTOR B (Wind Generation Facility) UNIT AMOUNT	CONTRACTOR B BID	WBS Total Bid Cost
1	1.1.1.1	Establish Benchmarks	1				$ -		$ -		$ -	$ -
		Surveyor	4	HR	150	10	$ 6,000.00		$ -		$ -	$ -
		Laborer	8	HR	60	10	$ 4,800.00		$ -		$ -	$ 10,800.00
											Total	
2	1.1.2	Excavation					$ -		$ -		$ -	$ -
	1.1.2.1	Soil Removal	1				$ -		$ -		$ -	$ -
		Excavators	4	EQ	125	24	$ 12,000.00		$ -		$ -	$ -
		Operator	6	HR	70	24	$ 10,080.00		$ -		$ -	$ -
		Laborers	15	HR	60	24	$ 21,600.00		$ -		$ -	$ 43,680.00
											Total	
							$ -		$ -		$ -	$ -
							$ -		$ -		$ -	$ -
							$ -		$ -		$ -	$ -
							$ -		$ -		$ -	$ -
							$ -		$ -		$ -	$ -
							$ -		$ -		$ -	$ -

Unit Categories:
HR - Human Resources
MAT - Material
EQ - Equipment

Figure 7-9 Bid Tabulation Form for 50+ for All Seasons

be the case in this situation because the company has no experience building a Solar Panel or Wind Turbine Power-Generation facility. This can take a long time in most companies unless the senior management of the company gets involved to accelerate the process.

2. After three subcontractors are approved, the Project Manager can contact each one for a bid proposal on the work, equipment, and material required for the installation of a Solar Panel Electric-Generation facility (Project No. 1) and a Wind Turbine Electric-Generation facility (Project No. 2).

3. The bids from each subcontractor must be carefully evaluated because some subcontractors usually provide a very low bid with the intent of obtaining a contract for the work, but they typically run out of money before the work is completed, which ends up becoming a major problem for the EPC Company.

To ensure that the required bid information would be consistent from each of the four projects, the Bid Tabulation Form, as shown in Figure 7-9, was provided to each of the four Project Managers for their project teams to utilize in compiling the bid information.

The reasons for the WBS number is in this form, which is critical for internal alignment and governance of the four different projects, are the following:

1. It ensures that there is no duplication of cost for each work activity and provides for a much quicker analysis and change of a deliverable's overall cost, if required.

2. It provides for the alignment of each project's scope to the requirements of the overall contractual workscope for the complete project and the EPC Company's vision for this project, which will be their first "green" type of complex construction project.

3. It supports a much quicker production of the required Critical Path Schedule (CPM) that the customer can quickly approve because all of the required durations for each of the activities are provided, along with their summation, at the higher levels of the WBS.

4. It provides a detailed cost breakdown for the Project Manager to use for the overall Retirement Community Beta Center project, which can be easily developed into an accurate project budget once the bid is accepted by the customer.

The advantage of No. 3 becomes more apparent when the WBS, shown in Figure 7-8, is transformed into a Gantt Chart, as shown in Figure 7-10.

Branch Name	Duration	Start	End	Priority	Completion	Resources	Work	Cost
Retirement Community Beta Project	1 day?	5/29/2017	5/29/2017	500	0%		0 hrs	$0.00
⊟ 1. Project No. 1 Solar- Generating Facility (Contractor)	1 day?	5/29/2017	5/29/2017	500	0%		0 hrs	$0.00
⊟ 1.1 Civil	1 day?	5/29/2017	5/29/2017	500	0%		0 hrs	$0.00
⊟ 1.1.1 Site Work	1 day?	5/29/2017	5/29/2017	500	0%		0 hrs	$0.00
1.1.1.1 Establish Benchmarks	1 day?	5/29/2017	5/29/2017	500	0%		0 hrs	$0.00
⊟ 1.1.2 Excavation	1 day?	5/29/2017	5/29/2017	500	0%		0 hrs	$0.00
1.1.2.1 Soil Removal	1 day?	5/29/2017	5/29/2017	500	0%		0 hrs	$0.00
1.1.3 Foundation	1 day?	5/29/2017	5/29/2017	500	0%		0 hrs	$0.00
1.2 Mechanical	1 day?	5/29/2017	5/29/2017	500	0%		0 hrs	$0.00
1.3 Electrical	1 day?	5/29/2017	5/29/2017	500	0%		0 hrs	$0.00
⊟ 2. Project No. 2 Wind- Generating Facility (Contractor)	1 day?	5/29/2017	5/29/2017	500	0%		0 hrs	$0.00
2.1 Civil	1 day?	5/29/2017	5/29/2017	500	0%		0 hrs	$0.00
2.2 Mechanical	1 day?	5/29/2017	5/29/2017	500	0%		0 hrs	$0.00
2.3 Electrical	1 day?	5/29/2017	5/29/2017	500	0%		0 hrs	$0.00
⊟ 3. Project No. 3 Transformer and Substation	1 day?	5/29/2017	5/29/2017	500	0%		0 hrs	$0.00
⊟ 3.1 Transformer	1 day?	5/29/2017	5/29/2017	500	0%		0 hrs	$0.00
3.1.1 Electrical	1 day?	5/29/2017	5/29/2017	500	0%		0 hrs	$0.00
3.1.2 Mechanical	1 day?	5/29/2017	5/29/2017	500	0%		0 hrs	$0.00
3.1.3 Civil	1 day?	5/29/2017	5/29/2017	500	0%		0 hrs	$0.00
⊟ 3.2 Substation	1 day?	5/29/2017	5/29/2017	500	0%		0 hrs	$0.00
3.2.1 Electrical	1 day?	5/29/2017	5/29/2017	500	0%		0 hrs	$0.00
3.2.2 Mechanical	1 day?	5/29/2017	5/29/2017	500	0%		0 hrs	$0.00
3.2.3 Civil	1 day?	5/29/2017	5/29/2017	500	0%		0 hrs	$0.00
⊟ 4. Project No. 4 Beta Retirement Community	1 day?	5/29/2017	5/29/2017	500	0%		0 hrs	$0.00
⊟ 4.1 Assisted Living Center	1 day?	5/29/2017	5/29/2017	500	0%		0 hrs	$0.00
4.1.1 Electrical	1 day?	5/29/2017	5/29/2017	500	0%		0 hrs	$0.00
4.1.2 Mechanical	1 day?	5/29/2017	5/29/2017	500	0%		0 hrs	$0.00
4.1.3 Civil	1 day?	5/29/2017	5/29/2017	500	0%		0 hrs	$0.00
⊟ 4.2 Senior Apartment Building	1 day?	5/29/2017	5/29/2017	500	0%		0 hrs	$0.00
4.2.1 Electrical	1 day?	5/29/2017	5/29/2017	500	0%		0 hrs	$0.00
4.2.2 Mechanical	1 day?	5/29/2017	5/29/2017	500	0%		0 hrs	$0.00
4.2.3 Civil	1 day?	5/29/2017	5/29/2017	500	0%		0 hrs	$0.00
⊟ 4.3 Sports Complex	1 day?	5/29/2017	5/29/2017	500	0%		0 hrs	$0.00
4.3.1 Electrical	1 day?	5/29/2017	5/29/2017	500	0%		0 hrs	$0.00
4.3.2 Mechanical	1 day?	5/29/2017	5/29/2017	500	0%		0 hrs	$0.00
4.3.3 Civil	1 day?	5/29/2017	5/29/2017	500	0%		0 hrs	$0.00
⊟ 4.4 Medical Clinic	1 day?	5/29/2017	5/29/2017	500	0%		0 hrs	$0.00
4.4.1 Electrical	1 day?	5/29/2017	5/29/2017	500	0%		0 hrs	$0.00
4.4.2 Mechanical	1 day?	5/29/2017	5/29/2017	500	0%		0 hrs	$0.00
4.4.3 Civil	1 day?	5/29/2017	5/29/2017	500	0%		0 hrs	$0.00

Figure 7-10   Gantt Chart for the Retirement Community Beta Project Work Breakdown Structure (WBS)

It can be seen how the Gantt Chart for the Retirement Community Beta will change as the cost and duration information for each activity are added to each project, which the company's Project Control team does as the bid compilation process moves forward to the finalization of the overall bid for this project. Figure 7-11 shows how this Gantt Chart (shown in Figure7-10) can be easily transformed into a required CPM.

Once the costs and durations for all of the activities are completed for each project, the next step will be to add the dates for each activity into the Gantt Chart shown in Figure 7-10 and the CPM Schedule shown in Figure 7-11, which must be adjusted, as required, to ensure that the contractual milestone dates for each project are met by the EPC Company.

	Task Mode	Task Name	Duration	Start	End	Pred...	Comple...	Priority	Resour...	Work	Cost
1		Retirement Community Beta Proj...	128.6 days ?	4/30/2018	12/10/2018		0%	500		0 hrs	$0.00
2		1. Project No. 1 Solar-Generatin...	128.6 days ?	4/30/2018	12/10/2018		0%	500		0 hrs	$0.00
3		1.1 Civil	128.6 days ?	4/30/2018	12/10/2018		0%	500		0 hrs	$0.00
4		1.1.1 Site Work	35.8 days	4/30/2018	6/29/2018		0%	500		0 hrs	$0.00
5		1.1.1.1 Establish Benchmarks	35.8 days	4/30/2018	6/29/2018		0%	500		0 hrs	$0.00
6		1.1.2 Excavation	92.8 days	6/29/2018	12/10/2018		0%	500		0 hrs	$0.00
7		1.1.2.1 Soil Removal	92.8 days	6/29/2018	12/10/2018	5	0%	500		0 hrs	$0.00
8		1.1.3 Foundation	0.8 day?	4/30/2018	4/30/2018		0%	500		0 hrs	$0.00
9		1.2 Mechanical	0.8 day?	4/30/2018	4/30/2018		0%	500		0 hrs	$0.00
10		1.3 Electrical	0.8 day?	4/30/2018	4/30/2018		0%	500		0 hrs	$0.00
11		2. Project No. 2 Wind-Generatin...	0.8 day?	4/30/2018	4/30/2018		0%	500		0 hrs	$0.00
12		2.1 Civil	0.8 day?	4/30/2018	4/30/2018		0%	500		0 hrs	$0.00
13		2.2 Mechanical	0.8 day?	4/30/2018	4/30/2018		0%	500		0 hrs	$0.00
14		2.3 Electrical	0.8 day?	4/30/2018	4/30/2018		0%	500		0 hrs	$0.00
15		3. Project No. 3 Transformer an...	0.8 day?	4/30/2018	4/30/2018		0%	500		0 hrs	$0.00
16		3.1 Transformer	0.8 day?	4/30/2018	4/30/2018		0%	500		0 hrs	$0.00
17		3.1.1 Electrical	0.8 day?	4/30/2018	4/30/2018		0%	500		0 hrs	$0.00
18		3.1.2 Mechanical	0.8 day?	4/30/2018	4/30/2018		0%	500		0 hrs	$0.00
19		3.1.3 Civil	0.8 day?	4/30/2018	4/30/2018		0%	500		0 hrs	$0.00
20		3.2 Substation	0.8 day?	4/30/2018	4/30/2018		0%	500		0 hrs	$0.00
21		3.2.1 Electrical	0.8 day?	4/30/2018	4/30/2018		0%	500		0 hrs	$0.00
22		3.2.2 Mechanical	0.8 day?	4/30/2018	4/30/2018		0%	500		0 hrs	$0.00
23		3.2.3 Civil	0.8 day?	4/30/2018	4/30/2018		0%	500		0 hrs	$0.00
24		4. Project No. 4 Beta Retiremen...	0.8 day?	4/30/2018	4/30/2018		0%	500		0 hrs	$0.00
25		4.1 Assisted Living Center	0.8 day?	4/30/2018	4/30/2018		0%	500		0 hrs	$0.00
26		4.1.1 Electrical	0.8 day?	4/30/2018	4/30/2018		0%	500		0 hrs	$0.00
27		4.1.2 Mechanical	0.8 day?	4/30/2018	4/30/2018		0%	500		0 hrs	$0.00
28		4.1.3 Civil	0.8 day?	4/30/2018	4/30/2018		0%	500		0 hrs	$0.00
29		4.2 Senior Apartment Building	0.8 day?	4/30/2018	4/30/2018		0%	500		0 hrs	$0.00
30		4.2.1 Electrical	0.8 day?	4/30/2018	4/30/2018		0%	500		0 hrs	$0.00
31		4.2.2 Mechanical	0.8 day?	4/30/2018	4/30/2018		0%	500		0 hrs	$0.00
32		4.2.3 Civil	0.8 day?	4/30/2018	4/30/2018		0%	500		0 hrs	$0.00
33		4.3 Sports Complex	0.8 day?	4/30/2018	4/30/2018		0%	500		0 hrs	$0.00
34		4.3.1 Electrical	0.8 day?	4/30/2018	4/30/2018		0%	500		0 hrs	$0.00
35		4.3.2 Mechanical	0.8 day?	4/30/2018	4/30/2018		0%	500		0 hrs	$0.00
36		4.3.3 Civil	0.8 day?	4/30/2018	4/30/2018		0%	500		0 hrs	$0.00
37		4.4 Medical Clinic	0.8 day?	4/30/2018	4/30/2018		0%	500		0 hrs	$0.00
38		4.4.1 Electrical	0.8 day?	4/30/2018	4/30/2018		0%	500		0 hrs	$0.00
39		4.4.2 Mechanical	0.8 day?	4/30/2018	4/30/2018		0%	500		0 hrs	$0.00
40		4.4.3 Civil	0.8 day?	4/30/2018	4/30/2018		0%	500		0 hrs	$0.00

Figure 7-11   Critical Path Schedule (CPM) Outline for the Retirement Community Beta Project

Project Presentation to Management

On December 5, 2017, the Project Manager, Bob, held a meeting with the Bid Compilation team to update the Bid Compilation Responsibility and Progress Checklist, shown in Table 7-2, and was satisfied with the results that were recorded, which are shown in Table 7-4.

The Program Manager, Sandy Xandau, and the Project Manager, Bob, had a meeting after this progress meeting to discuss their approach to the company's management for the Retirement Community Beta Project presentation, and they felt it would be best to bring the four Project Managers in at this time for the meeting, which would permit them to meet the company's sponsor and the other managers who will be overseeing this project. The other advantage is that each of these four Project Managers, who had already been involved with their respective projects in the bid compilation process, would be able to take immediate ownership if the company succeeds in being awarded the project by 50+ for All Seasons. The construction organizational chart that was developed, which is shown in Figure 7-12, represents the actual organization that will be established onsite to manage all of the construction work for the EPC Company.

The only position not filled in the organizational chart—that of the Senior Project Engineer for Project No. 4—will be filled by Nelson Summeri, a very experienced Senior Project Engineer with a Civil Engineering background, once another project for the EPC Company is completed (within a couple of weeks). The Program Manager, Sandy Xandau, would be the only person who will not be onsite all of the time because she will be busy ensuring that alignment and governance of the overall project is being maintained at both the site and corporate levels of the EPC Company, which is required to ensure success for this large, complex construction project.

On December 10th, the Project Manager, Bob, was notified that all of the required documents, as shown in Table No. 7-4, were now complete, which permitted him to finish the Retirement Community Beta Project presentation in time to meet with the Senior Vice President of Construction the following day for a "practice run" and to discuss any items that the Project Sponsor felt should be added prior to the final presentation to the EPC Company's senior management on December 14, 2017.

On December 14th, the final presentation of the Retirement Community Beta Project was conducted by the Project Manager, Bob, and the Program Manager, Sandy, and both received positive feedback on each of the items along with thanks for a "job well done" from the EPC Company's CEO, which was shared with the rest of the Bid Compilation team. Over the next few days, all of the remaining work required to ensure that the format of the bid proposal was in accordance with that specified by 50+ for All Seasons was completed and reviewed by each department head. After this, the Commercial Manager, the Senior Vice President of Sales and Marketing, and the Project Manager, Bob, traveled to the headquarters of 50+ for All Seasons and officially submitted the Comet EPC Company's bid for the Retirement Community Beta Project to their Program Manager at 11:00 AM on December 18, 2017.

7.1.2 Project Planning

On January 15, 2018, the Comet EPC Company received official notification that the 50+ for All Seasons was awarding them the contract for construction of this Retirement Community, which will be located approximately 20 miles NE of Tempe, Arizona. The Commercial Manager, the Senior Vice President of Sales and Marketing, and the Project Manager, Bob, were all present in the Senior Vice President of Construction's office as the Project Sponsor opened the package and read the award letter to the Retirement Community Beta Project Team.

After everyone was congratulated for this major achievement, which the Program Manager, Sandy, stated would be the first of many environmental projects for the company, the next immediate order of business was going to be a one-week Project Initiation phase gate review with all members of the Bid Compilation team and the Retirement Community Beta Project Team. The purpose of this one-week meeting, which will be held at local, popular resort on January 29th, is to not only capture what went "right" during the compilation process but also capture what went "wrong," so that when the next opportunity arises, the overall process of bid compilation for the company will be more effective and expedient. The other purpose of this phase gate review is to leverage the knowledge learned from certain areas to improve the Planning Phase for this complex construction project. For example, during the scope breakdown, there was some confusion over the interface areas between the engineering disciplines, such as civil foundations that required electrical conduit and earthing or grounding cabling, which slowed down this process because an Electrical Engineer had to be found that could define the scope of this electrical work and when it had to be installed.

The other important item expected to be completed in this meeting was an update of the Retirement Community Beta Project CEP that included the newly gained information for the construction of a solar electric-generating facility and a wind electric-generation facility, which will be updated even further during the Execution phase when construction work on Project Nos. 1 and 2 begins.

The new customer, 50+ for All Seasons, also stated in the award letter that they wanted to have a three-day project kick-off meeting in Tempe, Arizona,

Table 7-4 Bid Compilation Progress as of December 5, 2017

Project Management System Input	Department or Departments Responsible	Start Date	End Date	Current Progress (%)	Comments
Project Contract (Scope)	• Engineering Dept. Manager • Commercial Dept. Manager • Legal Dept. Manager • Construction Dept. Manager • Project Manager	10/17/17	10/30/17	100%	Completed as scheduled.
Project-Specific WBS	• Project Manager • Engineering Dept. Manager • Construction Dept. Manager • Project Engineer • Commercial Dept. Manager	10/31/17	11/21/17	100%	Subcontractors for Projects 1 and 2 pushed this out by 5 days to 11/26/27. The Construction Execution Plan (CEP) assisted keeping this activity on schedule.
Project Budget	• Accounting Dept. Manager • Construction Dept. Manager • Engineering Dept. Manager	11/22/17	12/9/17	95%	The bids from each of the contractors have been received.
Project Stakeholder List	• Project Manager • Commercial Dept. Manager • Project Engineer • Legal Dept. Manager	12/01/17	12/07/17	97%	The information from the Project 1 and 2 subcontractors is being added to this list.
Project Schedule (Level 4 or 5)	• Project Manager • Engineering Dept. Manager • Construction Dept. Manager • Project Engineer • Commercial Dept. Manager	11/22/17	12/9/17	92%	A Level 4 Project Schedule (CPM) will be provided as per the bid requirements from the customer.
Project Resource Requirements	• Project Manager • Engineering Dept. Manager • Construction Dept. Manager • Project Engineer • Commercial Dept. Manager	11/22/17	11/30/17	100%	The Project 1 and Project 2 subcontractors have been given a separate resource assignment, which pushed this out by 2 days.

(Continued on following page)

Table 7-4 Bid Compilation Progress as of December 5, 2017 (*Continued*)

Project Management System Input	Department or Departments Responsible	Start Date	End Date	Current Progress (%)	Comments
Project Engineering Requirements	• Project Manager • Engineering Dept. Manager • Construction Dept. Manager	11/22/17	11/28/17	100%	The Engineering requirements were reduced by the addition of the subcontractor's Engineering for Project Nos. 1 and 2.
Project Risk Register	• Project Manager • Engineering Dept. Manager • Construction Dept. Manager • Project Engineer • Commercial Dept. Manager	10/31/17	12/08/17	90%	The additional risk of the subcontractors for Project Nos. 1 and 2 is currently being evaluated.
Project HSE Requirements	• HSE Dept. Manager • Commercial Dept. Manager • Project Manager • Project Engineer	12/10/17	12/4/17	100%	The HSE Department was able to pull this activity back by 12 days to 12/4/17 due to historical information and assistance from the subcontractors for Project Nos. 1 and 2.

starting on February 20, 2018, and were looking forward to meeting the new Comet EPC Retirement Community Beta Project Team, which left the Project Manager, Bob, and the project team about three weeks to prepare for this kick-off meeting.

50+ for All Seasons Project Kick-Off Meeting

The Project Manager, Bob, was greeted by the Program Manager for 50+ for All Seasons, who had conducted the bid meeting, when he and the Comet EPC Construction Project team arrived for the project kick-off meeting. After everybody was introduced, the 50+ Program Manager first discussed what problems they had experienced with past projects, as follows:

1. **Poor Group Dynamics** – During construction of a large community complex with renewable sources of energy, problems will arise that will require both the customer's and the EPC Company's Site Project Team

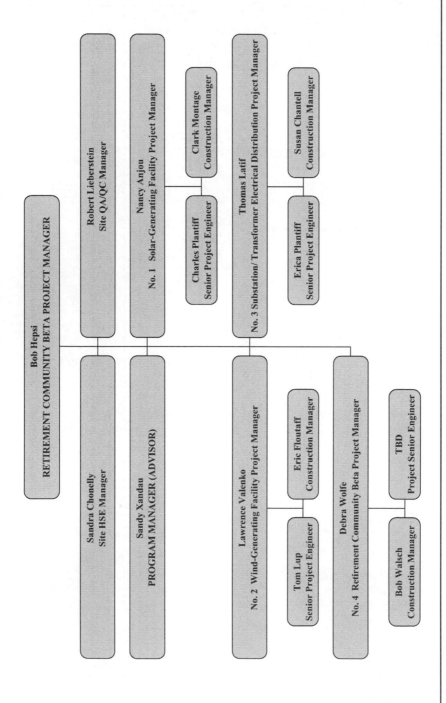

Figure 7-12 Comet EPC Four Project Organizational Chart for the Retirement Community Beta Construction Project

to work closely together for quick and efficient resolution. In these situations, 50+ has found that the tendency is for the Site Project Team to first take this problem offsite to their management before they discuss it with the Customer or the EPC Company, which only added to the complexity of the resolution process. To avoid these types of situations, 50+ has established a mandatory group workshop for all contractor Site Project Teams and the assigned 50+' Site Project Team, which will be a three-day team-building exercise that will be provided by a consultant that 50+ has hired for this purpose and will be conducted one week after 50+ receives the "Notice to Proceed."

2. **Transparency** – Another problem that 50+ has experienced in the past with other projects using EPC Contractors is a lack of transparency when it comes to onsite problems, health or safety issues, and project delays. To prevent this, the 50+ Site Project Manager, Roger Rejalf, will have a 30-minute site tour with the Comet EPC Project Manager, Bob, three times a week to evaluate site conditions and progress. In addition to this, 50+ will require that any safety issue, first aid or near-miss safety event, or environmental issue be reported to the 50+ Site HSE Manager within 24 hours. This is in the 50+ HSE Manual that has been handed out to everyone in this meeting.

3. **Interface** – On this project there will be four major external interface points, which the 50+ Site Project Manager, Roger, and his team will manage. These four external interface groups are the following;
   - **Maricopa County Utilities** – Sewage and potable water tie-ins.
   - **Maricopa County Highway Department** – County road access and shipments for the project.
   - **Arizona State Highway Department** – Access to state highways and shipment of large loads to the site.
   - **Arizona Public Service** – Utility tie-in to the substation and overhead power lines.

After the 50+ Program Manager was finished, the Comet EPC Project Manager, Bob, presented Comet's Site Organizational Chart, as shown in Figure 7-13.

The Comet EPC Program Manager, Sandy, explained that after evaluating the complexity of this particular project, a Site Engineering Manager was added, who will work with the corporate Engineering Manager to ensure that the design engineering for the project will remain aligned with the construction project's schedule and provide support for the Site QA/QC Department during the procurement and construction phase of this project. Sandy went on to explain that this complexity is also the reason that the customer sees four Project Managers working under the Site Project Manager, Bob, because there will be four individual projects, as shown in Table 7-5.

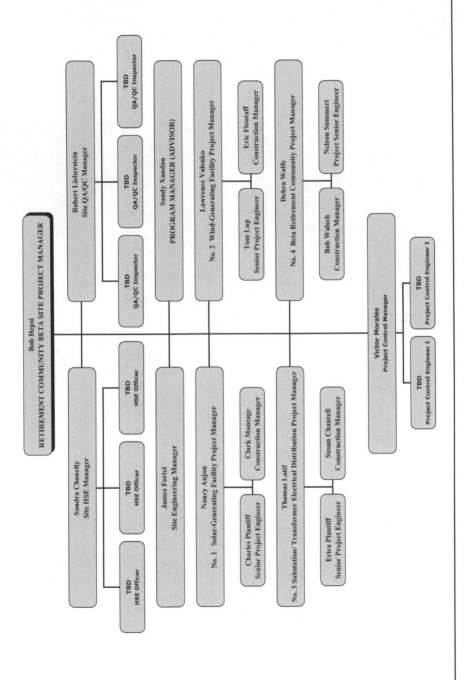

Figure 7-13 Comet EPC Retirement Community Beta Project Construction Organization

Table 7-5 Four Projects of the Retirement Community
Beta Construction Project

Project Number	Site Location	Project Manager
1	Solar Energy Electrical-Generating Area	Nancy Anjou
2	Wind Energy Electrical-Generating Area	Lawrence Valenko
3	Electric Distribution and Substation	Thomas Latif
4	Retirement Community Beta Retirement Community	Debra Wolfe

This approach will provide a larger amount of governance and greater alignment with both the overall project goal and the corporate goals for these types of projects within the Comet EPC Company.

The 50+ for All Seasons Site Project Manager, Roger, told Sandy that he was pleasantly surprised with this approach because it provides greater alignment with the project and corporate goals. This was something, he said, that 50+ for All Seasons found in past projects, which is why their project organization structure, as shown in Figure 7-14, is similar to that of Comet's site organization.

Another different approach that 50+ for All Seasons takes on their projects (see Figure 7-13) is that one project has a pool of project Civil Engineers, which is used for all three projects. This approach was taken because the Civil Engineering design and work for all of the three projects is primarily the same, with the exception being specific project-related requirements.

On the last day of the kick-off meeting, the Comet Project Sponsor, the Senior Vice President of Construction arrived for the official contract signing with the 50+ for All Seasons Project Sponsor, which was followed by a luncheon for all of the members of both companies that would be involved in the construction of this Retirement Community as well as local government officials who would also be involved in supporting the project. It was critical that all of the project stakeholders had a chance to meet face-to-face in this relaxed environment to voice their concerns with the management of both companies and register their support for the success of this project, which would the first of its kind for this area near Tempe, Arizona. At this signing of this contract, the 50+ for All Seasons Project Sponsor announced that they planned to receive the official "Notice to Proceed" by May 7, 2018.

Construction Initial Planning (1st Six Months)

After Project Manager, Bob, returned to the office, he set up a workshop with the Site Project Managers and Construction Managers to develop a six-month

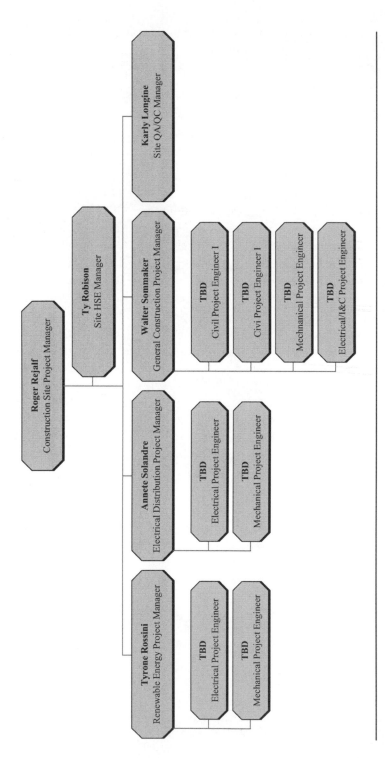

Figure 7-14 50+ for All Seasons Site Organizational Chart

look-ahead timeline and schedule, using February 23, 2018, the signing of the project's contract, as the first milestone, followed by May 7, 2018, which is the customer's expected official "Notice to Proceed" date. The result of this workshop was a six-month timeline, as shown in Figure 7-15.

The Gantt Chart that was the basis for this timeline (shown in Figure 7-16) indicated that the Comet EPC Company should start excavation of Project No. 1 on July 5, 2018, and have the construction staff moved into their temporary site offices by July 14, 2018.

After this schedule was reviewed and accepted by the Site Project Manager and construction managers, the Project Manager held a meeting with the two companies that would be installing the Solar Panel Electric-Generating facility and the Wind Turbine Electric-Generating facility, respectively, so that Nancy Anjou and Lawrence Valenko, the Comet Site Project Managers for Project Nos. 1 and 2, could explain the new six-month look-ahead schedule to their

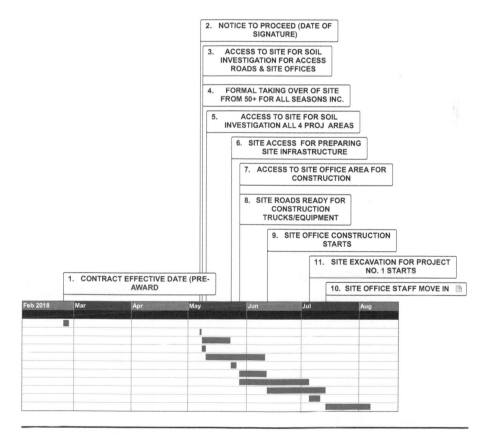

Figure 7-15 Retirement Community Beta Six-Month Look-Ahead Timeline

Retirement Community Beta Six-Month Look-Ahead Construction Gantt Chart

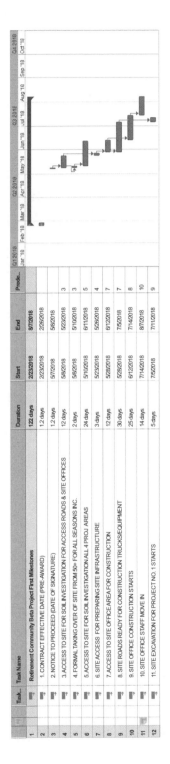

Task..	Task Name	Duration	Start	End	Prede..
1	Retirement Community Beta Project First Milestones	122 days	2/23/2018	8/7/2018	
2	1. CONTRACT EFFECTIVE DATE (PRE-AWARD)	1.2 days	2/23/2018	2/26/2018	
3	2. NOTICE TO PROCEED (DATE OF SIGNATURE)	1.2 days	5/7/2018	5/8/2018	
4	3. ACCESS TO SITE FOR SOIL INVESTIGATION FOR ACCESS ROADS & SITE OFFICES	12 days	5/8/2018	5/23/2018	3
5	4. FORMAL TAKING OVER OF SITE FROM 50+ FOR ALL SEASONS INC.	2 days	5/8/2018	5/10/2018	3
6	5. ACCESS TO SITE FOR SOIL INVESTIGATION ALL 4 PROJ AREAS	24 days	5/10/2018	6/11/2018	5
7	6. SITE ACCESS FOR PREPARING SITE INFRASTRUCTURE	3 days	5/23/2018	5/26/2018	4
8	7. ACCESS TO SITE OFFICE AREA FOR CONSTRUCTION	12 days	5/28/2018	6/12/2018	7
9	8. SITE ROADS READY FOR CONSTRUCTION TRUCKS/EQUIPMENT	30 days	5/28/2018	7/5/2018	7
10	9. SITE OFFICE CONSTRUCTION STARTS	25 days	6/12/2018	7/14/2018	8
11	10. SITE OFFICE STAFF MOVE IN	14 days	7/14/2018	8/7/2018	10
12	11. SITE EXCAVATION FOR PROJECT NO. 1 STARTS	5 days	7/5/2018	7/11/2018	9

Figure 7-16 Retirement Community Beta Project No. 1 (Solar-Generating Facility) Initial Construction Gantt Chart

respective subcontractor Project Managers and establish their mobilization plans because the Civil Work would be starting on these two areas in July before the other two areas.

This six-month timeline information also provided the Project Manager and the members of the Site Construction Project Team with an idea of how much time they had to finish the remaining components of the Project Management System, Work Management System, and Quality Management System, all of which had to be in operation prior to the start of construction activities onsite. The Program Manager, Sandy, stated the best way to ensure alignment and governance of these three systems among the four projects, as they are prepared for the construction work onsite, is as follows:

1. **Work Breakdown Structure (WBS Numbering)** – The WBS number that all schedules will use onsite and offsite should be in the format shown in Figure 7-17. This will permit each project to track their own progress and submit their weekly three-month look-ahead schedules, which will be the Level 5 type shown in Figure 7-17, to the Project Control Manager, Victor Morales, who will consolidate all four projects accordingly for reporting internally and to the customer. It will be the responsibility of each EPC Project Manager to ensure that their subcontractors are using the same WBS format when they submit their schedules each week. The CEP will also have the same WBS format applied to each of the activities so that the information for that activity can easily be placed into the project's schedule, which will expedite the production of a Level 5 schedule and permit project delays to be quickly analyzed by the Site Project Manager through the Project Management System for immediate corrective action.

2. **Communication Protocols** – There will be many stakeholders involved with this project, internally and externally, which mandates that all communication be identified correctly at the time it is written and before it is distributed. This identification will be performed through the usage of a protocol code for each document, which is shown in Figure 7-18.

   This format must be followed by all EPC Project Managers, project team members, and subcontractors when sending formal written communication, which includes cover sheets for Minutes of Meeting (MOM), site memos, site notifications, etc. The Document Control Center will provide the necessary number, which can be done online via access to the project's website. This will include transmittal of all documents to the Document Control Center for formal submittal. This same type of service extends to all project-related external stakeholders according to their communication preferences. However, formal communication

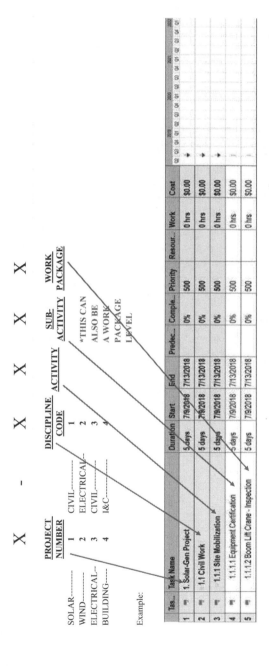

Figure 7-17 Work Breakdown Structure (WBS) Number Format for the Retirement Community Beta Construction Project

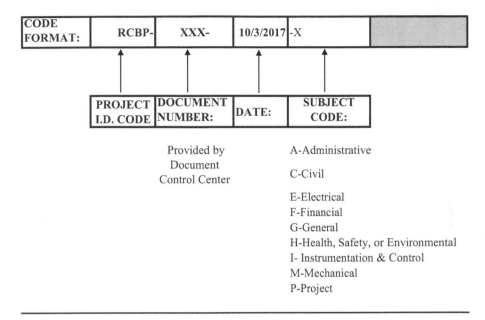

Figure 7-18   Retirement Community Beta Project Document Submittal Code Format

with external stakeholders will be a different process and involve specific access protocols, for security purposes. The Site QA/QC Department will perform an internal monthly communication audit to ensure that this particular protocol is being maintained onsite and offsite because all of the EPC corporate personnel, which includes management, are following this protocol when communicating with the site. The intent of this Communication Protocol policy is to ensure that all formal project communication is properly tracked throughout the organization at all levels and is easily retrieved because the Site Document Control Center will be handling and storing all formal communication.

Engineering and Quality Initial Planning

After these discussions, the Project Manager, Bob, contacted the Engineering Manager to discuss the onsite staffing requirements with James Fortel, the new Site Engineering Manager, because the amount of design packages, which include reviewing those from the Solar Panel and Wind Turbine installation contractors, will require the efforts of both onsite and offsite Engineering teams to get these reviewed quickly and over to the customer for final approval. All of them felt that reviewing this approval process, shown in Figure 7-19, would assist them in determining how many engineers would be required to get through these design packages, which will be in the thousands and will also include the finalization of a 3-D Model for each of the project's deliverables, which the customer must also approve.

The process flow chart clearly shows that the allocation of engineers would have to be shifted onto the civil side to provide more Civil Engineers onsite than planned. This is due to the two subcontractors needed for Project Nos. 1 and 2 along with the amount of engineering documentation that each will be required to submit for approval before they can mobilize to site. After their discussion, a table for engineering allocation and location was established, as shown in Table 7-6.

After these numbers were finalized and accepted by both managers, the Project Manager, Bob, updated the Project Engineer Plan in the Project Management System so that the additional cost and allocation of resources for the project could be evaluated and reported to the Project Sponsor. The original plan was that the company Engineering Staff would perform this work and charge the project's budget accordingly; however, the additional subcontractors, coupled with the site preparation work, changed that approach. After this was completed, the Project Manager, Bob, went through the baseline construction schedule so that both Engineering Managers could prioritize the drawings that would be required early in the project along with the respective engineering procedures, if required, and focus their resources to ensure that the engineering review and approval work would not impact the project's schedule. After this was completed, the two Engineering Managers and Bob met with Robert Lieberstein, the Site QA/QC Manager, to discuss what engineering support he required at this point in the project. Robert stated that he needs engineering support in three areas at this time, as follows:

1. **Local Concrete Supplier** – The QA/QC Department have identified two concrete suppliers in the local area that can provide the concrete for the project and are close enough so that the time from the supplier to the site is not an issue. However, they need one of the Site Civil Engineers to verify if each of these suppliers can provide the grade of concrete that will be required onsite and the quantity required, as well as meet the slump (concrete consistency) requirements at time of pour. James told him that he knows a Site Civil Engineer from his last project who did an excellent job on just this and that he is available to start work immediately with the Site QA/QC Group on this.

2. **Vendor Qualification** – There will be a number of equipment and material vendors for this project that are new to both the EPC Company and

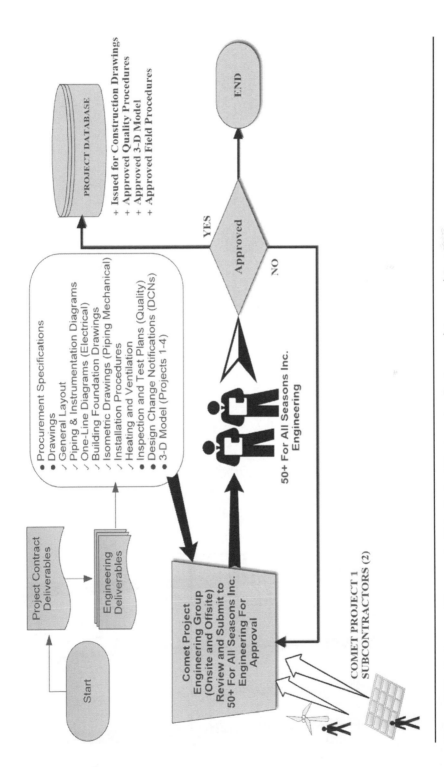

Figure 7-19 Retirement Community Beta Project Engineering Review and Approval Process

---

PROJECT DATABASE

+ **Issued for Construction Drawings**
+ **Approved Quality Procedures**
+ **Approved 3-D Model**
+ **Approved Field Procedures**

END

Approved

YES

NO

- Procurement Specifications
- Drawings
- General Layout
- Piping & Instrumentation Diagrams
- One-Line Diagrams (Electrical)
- Building Foundation Drawings
- Isometric Drawings (Piping Mechanical)
- Installation Procedures
- Heating and Ventilation
- Inspection and Test Plans (Quality)
- Design Change Notifications (DCNs)
- 3-D Model (Projects 1-4)

50+ For All Seasons Inc. Engineering

Project Contract Deliverables

Engineering Deliverables

Comet Project Engineering Group (Onsite and Offsite) Review and Submit to 50+ For All Seasons Inc. Engineering For Approval

Start

COMET PROJECT 1 SUBCONTRACTORS (2)

Table 7-6 Retirement Community Beta Construction Project Engineering Allocation

Project No.	Engineer Onsite (Discipline/Qty.)	Engineer Offsite (Discipline/Qty.)	Comments
1	Civil/2 Electrical/2 Mechanical/1	Civil/2 Electrical/2 Mechanical/1	The intent here is that the onsite Engineering staff will be dedicated to expediting the solar subcontractor's Engineering approval by providing preliminary reviews onsite.
2	Civil/1 Electrical/1 Mechanical/2	Civil/1 Electrical/1 Mechanical/2	The intent here is that the onsite Engineering staff will be dedicated to expediting the wind turbine subcontractor's Engineering approval by providing preliminary reviews onsite.
3	Electrical/1	Civil/2 Electrical/2 Mechanical/2	The electrical designs are similar to what the company has designed in the past, but the electrical designs will be different, so the site Electrical Engineer will review and report the Engineering changes that need to be approved.
4	Civil/3 Electrical/1 Mechanical/1	Civil/2 Electrical/2 Mechanical/2	The site Engineering staff will be involved with establishing the site infrastructure in addition to reviewing Engineering documents, with the priority being the site civil, electrical, and mechanical work.
Total Engineering Requirements	Civil/6 Electrical/5 Mechanical/4	Civil/7 Electrical/7 Mechanical/71	These requirements will be reevaluated once the Engineering Design Approval activities are completed.

the customer. They will, therefore, require support from Engineering to perform the required due diligence with the Site QA/QC Group to qualify them for this project prior to submitting their name and profile to the customer for their approval. The Engineering Manager told Robert that he already had a group of Engineers that worked with both the Procurement Department and the QA/QC Department on a previous project performing similar work, which he would assign to this project and make readily available to the Site QA/QC Group.

3. **Inspection and Test Plans (ITPs)** – The addition of two new subcontractors will significantly increase the number of factor ITPs, and Engineering Support will be required from each of the four projects to review these ITPs, approve them, and send them to the customer for their approval prior to the start of manufacturing. The priority at this time is the rebar supplier and the structural steel supplier, who have begun to outsource some of their material and components overseas, which is why the customer is not comfortable with the choice of these vendors for this project. The sooner that the QA/QC Group can confirm that each of these suppliers has a strong QA/QC program, with rigid quality standards and a Inspection Test Plan approved by the customer, the better it will be for the company and the overall project.

The Project Manager, Bob, stated that he would make item No. 3 the priority at this time, which had the agreement of both Engineering Managers. He would then work Item No. 2 and Item No. 1 in parallel with the QA/QC Group so that all vendors and the concrete supplier could be approved by the customer before the start of construction in July. This would be intense for both the Engineering Department, onsite and offsite, and the QA/QC Group, but possible because the structural steel and rebar suppliers were companies that Comet has worked with for the past ten years on various projects with no problems in the area of quality.

Preliminary Site Survey

James, the Site Engineering Manager, showed the Project Manager, Bob, a rough sketch of the site access roads for each of the four projects, which the Civil Engineering Team had put together (see Figure 7-20).

His concern was that the Civil Engineering Team needed to get started on its design now because they also had to evaluate where the site drainage trenches would be installed in each of the four areas. The critical areas shown in this sketch that need to be urgently clarified are the following:

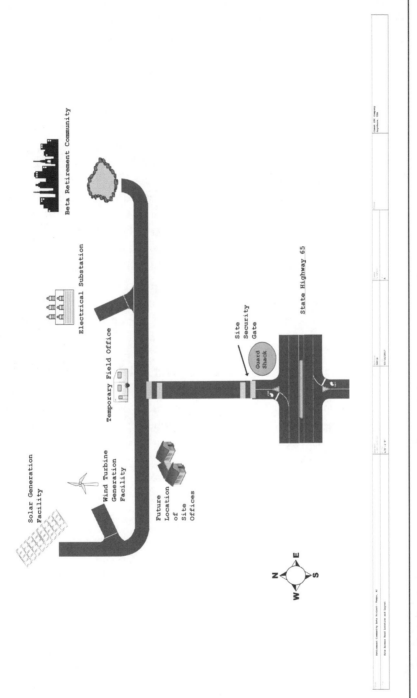

Figure 7-20 Retirement Community Beta Access Road Sketch

1. Point of access from Highway 65 and how this area will fenced, which must include some type of security building
2. The exact location of both the Temporary and Permanent Site Offices, which must include all building services
3. The condition of the ground and the area where the access roads are proposed to be installed
4. The best location for a drainage trench on both sides of these access roads so that all four areas will not be flooded in the event of a large rainfall

After this meeting was concluded, Bob immediately called the 50+ Site Project Manager, Roger, and asked him if it would be possible for the Site Engineering Manager, James, along with a few Civil Engineers, to have a preliminary site visit to establish the location of the site access roads for each of the four areas and the layout for the site drains. Roger replied that the county was supportive in this area and would probably approve this request because they were already modifying one area of Highway 65 for access to the construction site, but they would require an official letter from the Comet EPC Company specifying the purpose of this visit and the personnel who would be conducting this visit. Bob told Roger that he would have the letter on his desk tomorrow and that Comet appreciated his support on this. The following week, Roger called Bob and told him that the county had approved a Preliminary Site Visit for April 3, 2018, from 8:00 AM until 5:30 PM, with an extension of one day in the event that all of the work could not be completed in one day. Roger also stated that he would send over a copy of this authorization letter via email to Bob, but the official letter, which the Comet Civil team would need for the visit, should arrive tomorrow. After Bob received a copy of the county's authorization letter, he had James, the Site Engineering Manager, come to his office so that they could review it together and plan the trip for the Site Civil Team. James was surprised that the county, from the pictures attached, had almost completed the construction site access area, which was well in advance of the "Notice to Proceed" date of May 7, 2018.

On April 3, 2018, James and the Comet Civil Team were met by the county's Resident Engineer at the Highway 65 turnoff, which was about a quarter mile from the construction site. The Resident Engineer informed the James and his team that the county not only is a major stakeholder in this project, but that they want this project be a great success for the Comet EPC Company because they will be hiring a lot of local labor once the construction begins. James thanked the county's Resident Engineer for his support and asked if the county would be willing to provide Comet with the real-time X (Axial Direction), Y (Transverse Direction), and Z (Vertical Direction) benchmark positions, if available, of the following five interface points:

1. Highway 65 access point
2. County potable water supply point
3. County sewage system access point
4. County waste water system access point
5. County storm water system access point

Roger told the Resident Engineer that this would help Comet tremendously in establishing the correct benchmarks for their construction site 3-D Model, which was currently in progress and the reason for today's visit. The county's Resident Engineer told him that the county had just completed this type of survey a couple of weeks ago and would provide the required information to 50+ for All Seasons because they were the construction permit holder so that they could forward it to the Comet EPC Company. Roger was glad to hear this and placed it in the MOM for this Preliminary Site Visit along with the other items that the Civil Engineering Team provided as they completed their survey of the construction site.

The real-time X, Y, and Z benchmark positions from the county's Resident Engineer arrived attached to an official letter from 50+ for All Seasons. These positions, combined with the original site survey from the customer, permitted the Site Civil Engineering team to start development of the initial Site Civil Workscope, which would be required before construction work could start onsite.

Readiness Review

Because the 8th of May was less than one month away and the Preliminary Site Survey revealed that the county was farther along on the access road than previously thought, the Project Manager, Bob, informed the Site Construction Team and the Department Heads that he wanted to conduct an intense four-hour meeting on April 16, 2018. The purpose of this meeting will be a Readiness Review for the start of the project's Execution Phase, which will start at the time the site is turned over. It will determine the status of the planning work, which should be almost completed and should include the engineering design and review work and implementation of the project's Quality Management Plan. The meeting's agenda was laid out, including the three systems required to manage this large, complex construction project, and the Project Manager went down the list soliciting feedback on the major components for each of these systems. This provided him with confidence that the Project Team was ready to start the construction of the deliverables for this project. The agenda and responses were as follows:

A. Project Management System

1. **Project Quality Plan** – The Project Quality Plan was updated with information from the solar-generating facility and wind-generating facility contractors. After it was approved, it was sent over to the customer for their review and approval. The customer's approval was received officially via letter last week.

2. **Project Procurement Plan** – The Project Procurement Plan was reviewed by the company's senior management a few months ago and was approved in the middle of March. This item is now closed.

3. **Project Charter** – This item was closed one week after the contract was received on January 15, 2018, and the charter was updated accordingly.

4. **Project Communication Plan** – The Project Communication Plan was updated with the requirements of all of the stakeholders (along with their contact information), which included the two new contractors, and the latest Communication Protocols provided by the Program Manager, Sandy. The customer reviewed this latest version and approved it two week ago, which closed this item.

5. **Project Document Plan** – The Project Document Plan was updated to include the documents that will be provided by the solar-generating facility and wind-generating facility contractors. It was closed last month after each contractor confirmed that the information and format was correct.

6. **Project Staffing/Resource Plan** – The staffing for all four projects has been completed, which includes the resource requirements for the solar-generating facility and wind-generating facility contractors. It will be continually updated as this project moves into the Execution Phase and construction starts onsite.

7. **Project Logistics Plan** – The Logistics group is currently aligning the new Project Construction Schedule with the customer-approved equipment and material providers. This alignment includes reviewing what interstate permits will be required for the very large loads that will need a qualified heavy hauling company. It is being compiled at this time and will be done just after construction work starts onsite.

8. **Project Risk Management Plan** – The Project Risk Management Plan has been updated during the Planning phase with additional inputs from the solar-generating and wind-generating facility contractors and the risk of site turnover, which presents a large impact if it slides.

9. **Project Financial Register** – The Project Financial Register has been completed but will be consistently updated weekly once the site is turned over and construction work begins.

10. **Project Management Plan** – The Project Management Plan, with the assistance of the Program Manager, Sandy, has been updated with the new Communication Protocols, the new Organization Structure showing four projects under the umbrella of the Retirement Community Beta Project, and the new WBS structure that will be utilized by all stakeholders.

11. **Project HSE Plan** – The Project HSE Plan has been completed after being updated with the unique HSE information from the solar-generating and wind-generating facility contractors.

12. **Project Construction Execution Plan** – The CEP for the project was updated with the following information:
    a. The installation activities and their requirements with regard to resources, equipment, and materials for the installation of the solar power-generating facility
    b. The installation activities and their requirements with regard to resources, equipment, and materials for the installation of the wind-generating facility
    c. The finalized amount of access roads along with resources, equipment, and material from the site visit on April 3, 2018, which includes all of the required site drainage system
       - The CEP is completed now, but it will be updated as the construction work onsite progresses so that all "lessons learned" on this large, complex project will be effectively and accurately captured for future projects.

13. **Project Engineering Plan** – The Project Engineering Plan, as established in the Initiation Phase of this project, has increased in its scope, and the large number of drawings has been broken down into the following areas:
    a. **Computer 3-D Model** – The first 3-D Model, which is for the site arrangement of the four projects along with access roads and drains, has been updated with the county's benchmarks and submitted to the customer for their review and approval. The Site Engineering Manager, James, stated that he expected approval this week. The Mechanical, Electrical, and Civil Engineering groups have already started on the other 3-D Models for each of the four projects.
    b. **Tier I Drawings** – These drawings are all that will be required when the site work and excavations begin for the various foundations. This work is primarily Civil, in nature, but it also includes Electrical and Mechanical work, in some areas. At this time, the access road and drain civil drawings have been approved by the customer and will be issued in time to support the start of this work onsite on May 8, 2018. The Civil Engineering Team has already started submitting some of

the remaining drawings for all of the four projects to the customer for their review and approval. The target is to have all of the Tier 1 Drawings completed with customer approval and submitted to the Document Control Center by June 15, 2018.

The other Tier I drawings are the Piping and Instrumentation Diagrams (P&IDs), which are required for the systems within each of the 3-D Models. These are now being compiled in parallel to the 3D-Model development, and the combined Engineering team is targeting the completion of all P&IDs by June 30, 2018.

c. **Tier II Drawings** – These drawings involve providing the technical direction and quality inspections for the following activities:
  - Erection of the rebar foundation structures after lean concrete has been poured
  - Pouring of the concrete for of the completed rebar foundations
  - Installation of all the underground mechanical piping and electrical wiring, as required for each foundation
  - Installation of the grounding or earthing mat for each of the four project areas
  - Start of the excavation, compaction, and installation of the underground electrical duct banks or tunnels

A small contingent of the Engineering Office engineers has started reviewing some of these Tier II drawings as the respective 3-D Models are being completed.

d. **Tier III** – These drawings involve providing the technical direction for the next phase of construction , which is primarily Mechanical and Electrical, along with the respective quality procedures and inspections for the following activities:
  - Completion of the underground duct banks or tunnels for underground electrical cables and the cable pulling pit structures that provide cable access for installation
  - Erection of the various steel structures required for the installation of equipment, piping, or cable trays
  - Pouring of columns for various buildings, followed by the floors and roofs
  - Backfilling around the underground duct banks, trenches, and piping to final grade after compaction is completed
  - Erection of the Wind Turbine support structure and the support structure for the Solar Array

The Site and Engineering Office Engineers have prioritized the Tier III drawings and quality procedures so that their review and approval will be aligned to the current construction schedule.

e. **Tier IV** – These drawings involve providing the technical direction for the final construction activities, along with their respective quality procedures and inspections, for the following activities:
- Installation of all electrical panels followed by the pulling of electrical cables and termination of each cable in these panels
- Installation of the piping for the underground services, such as firefighting, drains, potable water, etc.
- Installation of the Wind Turbine along with all associated wiring and auxiliary support systems
- Installation of the Solar Panels along with all associated wiring
- Installation of the electrical wiring, plumbing, and both plumbing and electrical fixtures in each of the four buildings for the Retirement Community Beta Complex
- Pulling of the high-voltage cables from the Solar- and Wind Power-Generating Facilities to the Transformers and the Substation

After the Site and Engineering Office Engineers are finished with the Tier III drawings and quality procedures, they will work with all subcontractors to compile these Tier IV drawings and quality procedures, which will then be uploaded into the Master Drawing List (MDL), after the customer has approved each of them.

This approach will permit both Engineering Managers to effectively use their pool of engineers and, at the same time, accelerate the completion of all drawings, which will be in alignment with the current construction schedule and include the required quality documentation.

B. Work Management System – The Solar Power-Generating Facility Construction Manager, Clark Montage, stated that the four Project Construction Managers reviewed the Project's CEP along with the Baseline Construction Schedule to determine the best way to quickly finish the site roads and drains. This review provided an action plan, which began two weeks ago, outlined below:

1. **Soil Investigation** – The initial site survey analysis performed on April 3, 2018, permitted the Construction Management and Civil Engineering Team to develop a soil survey map for the access roads, site office areas (temporary and permanent), and all four project areas. To accelerate this work starting May 8, 2018, the Project Manager, Bob, and Construction Management team decided to utilize two contractors instead of one, as originally planned. This will increase the cost of this work by 10%, but it will reduce the current 12 days scheduled for road access and site office areas to just 5 days, which will permit the start of the construction work

on these areas to move from May 23, 2018, to May 14, 2018. This is shown in Activity No. 3 and Activity No. 6 in Figure 7-21.

2. **Site Construction Work** – To try to start the Civil Work on Project No. 1 a little earlier, the Activity No. 8 in Figure 7-21, which stands at 30 days, was evaluated with the assistance of the Construction Management team. The two customer-approved civil contractors were also contacted in this evaluation process to show how they could, with less than a 3% increase in cost, possibly reduce these 30 days by 5 to 10 days. The results of their efforts, combined with the assistance of the Construction Management team, provided the following information (shown in Table 7-7).

   An evaluation of the options provided in Table 7-7 reveal that just moving the working hours out by 2 hours is the most effective solution that will require only a small cost increase beyond the target of 3%. This internal change was presented to the company's senior management by the Project Manager, and it was approved. The Site HSE Manager, Sandra Chonelly, informed me at the end of last week that the company has plenty of portable lighting generators in the company's warehouse to support this activity. A letter regarding this change in the working hours for the initial Site Civil Work is already being drafted and will be issued to both the customer and the local county authorities by the end of next week.

3. **Fill Material and Gravel** – Three suppliers of fill material and gravel for all Civil Work have now been approved by the customer, and the Procurement Department is now completing the contract for each one of them to review and sign.

4. **Concrete Batching Plants** – Two concrete batching plants have now been approved by the customer after the tests of the concrete specified by the Civil Engineering Team have come back with positive results.

5. **Site Health, Safety, and Environmental Plan (HSE Plan)** – The Site HSE Manager, Sandra Chonelly, has contacted the customer and the local county authorities regarding establishing a perimeter fence around the whole project, which will include one separate gate for heavy vehicles and another for light vehicles. Their approval of Comet's security plan was received last month, and a contractor, with the permission of both parties, has started on the fencing installation, which is being supervised by the onsite Civil Engineering Team and should be completed by the end of May. This early action will ensure that this issue will not delay the "Notice to Proceed" scheduled for May 7, 2018.

C. Quality Management System – The Site QA/QC Manager, Robert Lieberstein, stated that in February, after the contract was awarded, the QA/QC Department got an early start on examining all of the documentation that

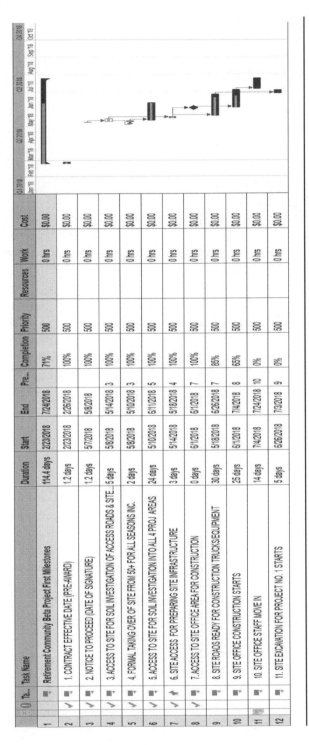

Figure 7-21 Retirement Community Beta Six-Month Seven-Day Improvement Construction Gantt Chart

Table 7-7 Retirement Community Beta Construction Project Site Access Road Civil Duration Improvement Analysis

Option	Contract Cost Increase (%)	Potential Time Savings (Days)	Remarks
1. Each Contractor Add 4 Trucks	6%	4	This would increase site traffic.
2. Add a Spoils Dumping Area on Site	4%	6	This will be handling material twice.
3. Each Contractor Add 2 Trucks and 1 Excavator	5%	8	This would increase site traffic, and specific areas will need to be prioritized.
4. Each Contractor Extend Working Hours to 12-Hour Work Days	3.5%	7	The customer and the local authorities have to be advised of this change. Temporary lighting will have to be provided for night work.

would be required for this project as well as on the timing of each document over the life of this project. This examination led to the following action items, which are either completed, in progress, or being developed at this time:

1. **Document Control** – The company's management, at the request of the QA/QC Department and the Program Manager, Sandy, decided to have a dedicated server for this project's massive amount of documentation and to provide an online QA/QC clearinghouse that both the EPC Company and customer can use for document review and approval. A kickoff meeting for this Document Control Program was conducted with the customer on February 16, 2018, and was followed by a five-day workshop on February 19, 2018. This workshop involved IT engineers, QA/QC Managers, Construction Managers, and Engineering Managers from both 50+ and the Comet EPC Company. This workshop permitted streamlining the process for all stakeholders so that the time between submittal and approval by the customer was reduced to just two weeks, which is why the Site Project Quality Plan has already been approved, along with the access road and drain civil drawings, by the customer. This new Document Control Program provides a very flexible QA/QC database that can quickly assimilate new documents, such as the QA/QC documents from the new solar power and wind power construction subcontractors.

2. **Method Statements (MSs)** – The Civil Engineering Team was able to work with the two subcontractors for the roads and the one for the drains

on detailed MSs for all of this work. These MSs were submitted to the customer for their review and approval, which the QA/QC Department expects to have before the start of work on May 14, 2015.

3. **Inspection and Test Plans (ITPs)** – The Engineering Team, onsite and offsite, has worked with the subcontractors and their suppliers to develop a factory Inspection and Test Plan (ITP) for all of the equipment that will be coming to the site. This effort is still in progress and will be concluded by the end of May. At this time, it has produced the following ITPs:

- Solar Panels and Associated Structure (Project No. 1)
- Wind Turbine and Generator (Project No. 2)
- Wind Turbine Base Plate and Structure (Project No. 2)
- High-Voltage Cable (Project No. 3)
- High-Voltage and Medium-Voltage Switchgear (Project No. 3)
- Heating, Ventilation, and Air Conditioning (HVAC) Equipment for all Buildings (Project No. 4)
- Medical Clinic Equipment
- Swimming Pool Filtration Equipment (Project No. 4)
- Structural Steel for all Buildings (Project No. 4)
- Modular Sections for the Permanent Site Office

The customer is now reviewing these ITPs in order of priority as established in the Project's Baseline Schedule, which should reduce the risk of any critical equipment being delivered late to the site. After these are received from the customer, the Procurement Department along with the Project Management team will establish a schedule for all of the customer inspections at each of the equipment supplier's location that will be in alignment with the Project's Baseline Schedule. This schedule will then be set up in the Document Control Program so the customer's QA/QC Manager will receive 30 days' advance notification of all factory inspections, which is per the contract.

After this meeting was completed, the Project Manager, Bob, and the Program Manager, Sandy, reviewed the MOM and determined that they could announce to the company's senior management that the Project Team was now ready to start the Execution Phase of this project. This announcement was given verbally to the company's senior management and the Project Sponsor in a brief one-hour presentation the following day, April 17, 2018.

7.1.3 Project Execution

On May 7, 2018, the 50+ for All Seasons Construction Site Project Manager, Roger Rejalf, officially received the "Notice to Proceed" and in a meeting onsite later that day, gave this to Bob Hepsi, Comet EPC Company's Senior Project Manager, so that construction work could start immediately onsite. This was then followed by a "ground-breaking" ceremony by the senior management of both companies, with the local Maricopa County government officials in attendance.

Soil Investigation Work

After the Comet HSE Manager completed the Site Safety Induction for all of the soil investigation subcontractor's personnel, the Project Manager, Bob, conducted a "kick-off" meeting with the subcontractor's Site Management Team. In this meeting, all four Project Managers for each of the four areas, along with Bob, discussed the subcontractor's schedule to ensure that their priorities were the same as Comet's Construction Baseline Schedule (shown in Table 7-8).

Table 7-8 Soil Investigation Schedule Area Priorities

Area	Start Date	Late Finish	Early Finish	Target Date
No. 1 (Solar Electric-Generating Plant)	5/10/2018	5/18/2018	5/16/2018	5/17/2018
No. 2 (Wind Electric-Generating Plant)	5/14/2018	5/22/2018	5/20/2018	5/21/2018
No. 3 (Electric Distribution and Substation)	5/17/2018	5/25/2018	5/23/2018	5/24/2018
No. 4 (Electric Distribution and Substation)	5/22/2018	5/30/2018	5/28/2018	5/29/2018

The subcontractor's Project Manager confirmed that their schedule was aligned to the priorities shown in Table 7-8, and he was confident that these dates would not be a problem to meet for his site team. After this, the Project Manager, Bob, introduced the Site Civil Engineering Team members along with the Construction Manager for each of the four areas so that the subcontractor's team members knew who their contacts were for this site work.

The soil investigation work began as scheduled on May 10, 2018, and the two-team approach of the subcontractor permitted the areas designed by the Site Civil Engineering Team to proceed quickly and smoothly. In Area No. 2, which is for the Wind Turbine Electrical-Generating Facility, the data provided by the subcontractor was much different from the other areas, and a review by the Site Civil Engineer revealed that the area for the foundation for the Wind Turbine tower may require soil improvement to improve its stability and loading capacity because it had higher levels of sand than the other areas. This finding and concern was brought to the No. 2 Project Site Construction Manager, Eric Floutaff, for his review and action, after his meetings with the company's Civil

Engineering staff and the Wind Turbine manufacturer's engineering group. The results of this review were that at least 20 piles would have to be installed prior to the start of the foundation work for the Wind Turbine structure. After this was confirmed, he filled out the necessary project change request, as shown in Figure 7-22.

Once this form was completed, he submitted the change request form to the No. 2 Wind-Generating Facility Project Manager, Lawrence Valenko, for his review and discussion with the Senior Project Manager, Bob, because he was the one who had to sign it. Bob called the Project Sponsor and informed him that this change request would require the Change Board's approval because it would be more than $50,000 and was not in the project's budget. The Project Sponsor, after reviewing this change request, contacted Bob and told him that he would ensure that the Board provides approval within seven working days. During these seven days, the Senior Project Manager, Bob, with the assistance of the No. 2 Project Manager and his team, took the following steps:

Step No. 1: They developed the Work Scope by first determining the size of the pile required, with the assistance of the Site Civil Engineering Team, and then completed the drawing that showed where and how they will be installed. After this was done, they complied with the Quality and Safety requirements for the installation of these piles, which the subcontractor would have to comply with after they arrived onsite and started work.

Step No. 2: They contacted the company's Procurement Department with the information from Step 1 above so that they could prepare the required bid specification and identify which subcontractors could perform this work, which the customer would approve.

Step No. 3: The Site QA/QC Manager contacted the subcontractor performing the site soil studies and asked them to compile their data for this change request and establish the Quality MS for the five test piles that Civil Engineering confirmed must be done immediately. The subcontractor's Project Manager confirmed that they had a company that could perform this test with the specified piles and had worked in Maricopa County before on other past projects. This information was passed over to the Senior Project Manager, Bob, so that he could get the company's and customer's permission to sole source this work, which would save the project money on mobilization costs and provide the results much more quickly for analysis.

Step No. 4: The No. 2 Site Construction Manager, with the approval of the other Construction Managers, changed the priority of the access road construction so that the road required for piling rig access to the Wind Turbine

β	Retirement Community Beta Project

	PROJECT CHANGE REQUEST			
Project:	Wind Electrical-Generating Facility (No. 2)		Date:	16-May-18
*Change Request Number: 2-001				
*Number Format: Project Number - Change Request No. (Provided by Document Control)				
Name and Title of Requestor: Eric Floutaff, Project No. 2 Site Construction Manager				
Project Change Category (Check Below All That Apply)				
X Schedule X Cost □ Scope □ Contract Requirements or Deliverables				
□ Quality/Additional Testing □ Additional Resources				
What Is Affected By This Change?				
X Construction □ Engineering Design □ Commissioning □ Equipment Delivery				
□ Contract □ Other				
Change Description Being Proposed:				
The foundation is for the Wind Turbine support structure will have to be changed to one that includes piles.				
What Is the Reason for the Change:				
The soil analysis revealed that it will be not stable enough to properly support this structure without significant soil improvement.				
What Are the Alternative Solutions (If Any):				
None				
What Work, Material, or Equipment Will Be Required for This Change:				
One Piling Rig, 20 spun piles (Diameter & Length to Be Determined by Civil Engineering) and a boom lift truck for delivery of the piles.				
What Additional Resources or Capital Will be Required for This Change:				
At least 5 Test Piles will first need to be conducted to confirm that spun pile selection is correct. After this successful test is complete, a Pile Integrity Test will have to be performed on all of the 20 piles installed. The overall cost for everything will be approximately $150,000 at this time (per Construction Execution Plan (CEP)).				
What Will Be the Risks Associated with This Change:				
Failure and Replacement of a Pile After Installation.				
What Will Be the Impact on Quality Related to This Change:				
Quality of the Wind Turbine Foundation will be improved and in accordance with the equipment manufacturer's specifications.				
Disposition:				
□ Approve □ Reject □ Defer				
Justification of Approval, Rejection, or Deferral:				
Senior Project Manager Signature:				

Comet Change Board Approval (Required for All Change Requests > $50K) :		
Name	Signature	Date

Figure 7-22   Project Change Request for Project No. 2 Piling Requirement

Foundation area would be completed first, which would permit this work to start in parallel to the installation of the site offices, which would start on June 2, 2018.

Step No. 5: After the scope was finalized, the No. 2 Site Construction Manager conducted a meeting with the Site Civil Engineering Team, the Site QA/QC Manager, and the No. 2 Project Site Construction Manager to compile an accurate cost for all of this work, which included the Test Piling and the Pile Integrity Testing, once the piles were installed. The complete cost breakdown for this first change request is shown in Table 7-9.

Table 7-9 Cost Breakdown for the No. 2 Project 20-Pile Installation on the Wind Turbine Foundation

Category	Cost (USD)	Cost (per Day)	Duration (Days)	Total Cost (USD)
Equipment (1 Piling Rig)		$5,000	5	$25,000
Material (20 Spun Piles)	$80,000			$80,000
Material (5 Test Piles)	$20,000			$20,000
Labor (7 Days – 10 People)		$4,250	7	$29,750
Services (Pile-Testing Team)		$2,500	3	$7,500
Total Cost for Change Request 2-001 (Piling of Wind Turbine Foundation)				$162,250

This final estimated figure was provided to the Senior Project Manager so that he could revise the amount provided to the Project Sponsor, The Project Sponsor will use this number as the basis for the company's Change Control Board's approval of $200,000 for this first change order because there may be some additional hidden costs not foreseen until the work starts.

After this was done, the Senior Project Manager, Bob, updated the project's Financial Register with the financial information and the project's CEP with the information developed in the first three steps. Bob also made sure that the Project Control team, with customer approval, updated the project's Baseline Schedule for Project No. 2 with the piling activities, as shown in Figure 7-23.

The Site QA/QC Manager, Robert Lieberstein, also made sure that the Project Quality Plan was updated with the quality program required for the piling work and requested that the piling subcontractor provide an MS for his QA/QC Engineers to review and approve prior to the start of work.

The company's Change Control Board approved this $200,000 piling change order on May 24, 2018, which permitted the piling subcontractor to be approved immediately and establish their mobilization date as May 28, 2018, so that they could support the test pile installation scheduled for June 6, 2018.

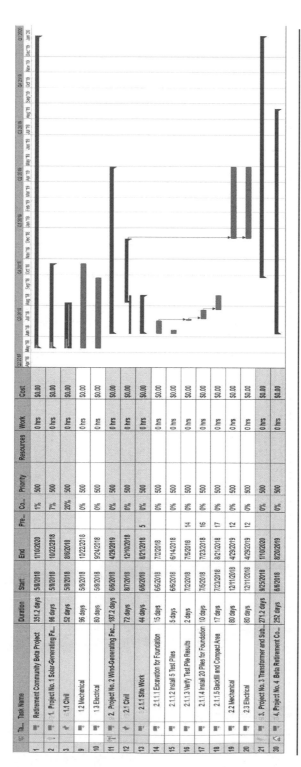

ID	Ta...	Task Name	Duration	Start	End	Pre...	Co...	Priority	Resources	Work	Cost
1		Retirement Community Beta Project	351.2 days	5/8/2018	1/10/2020		1%	500		0 hrs	$0.00
2		1. Project No. 1 Solar-Generating Fa...	96 days	5/8/2018	10/22/2018		7%	500		0 hrs	$0.00
3		1.1 Civil	52 days	5/8/2018	8/6/2018		20%	500		0 hrs	$0.00
9		1.2 Mechanical	96 days	5/8/2018	10/22/2018		0%	500		0 hrs	$0.00
10		1.3 Electrical	80 days	5/8/2018	9/24/2018		0%	500		0 hrs	$0.00
11		2. Project No. 2 Wind-Generating Fac...	187.2 days	6/6/2018	4/29/2019		0%	500		0 hrs	$0.00
12		2.1 Civil	72 days	8/7/2018	12/10/2018		0%	500		0 hrs	$0.00
13		2.1.1 Site Work	44 days	6/6/2018	8/21/2018	5	0%	500		0 hrs	$0.00
14		2.1.1.1 Excavation for Foundation	15 days	6/6/2018	7/2/2018		0%	500		0 hrs	$0.00
15		2.1.1.2 Install 5 Test Piles	5 days	6/6/2018	6/14/2018		0%	500		0 hrs	$0.00
16		2.1.1.3 Verify Test Pile Results	2 days	7/2/2018	7/5/2018	14	0%	500		0 hrs	$0.00
17		2.1.1.4 Install 20 Piles for Foundation	10 days	7/5/2018	7/23/2018	16	0%	500		0 hrs	$0.00
18		2.1.1.5 Backfill and Compact Area	17 days	7/23/2018	8/21/2018	17	0%	500		0 hrs	$0.00
19		2.2 Mechanical	80 days	12/11/2018	4/29/2019	12	0%	500		0 hrs	$0.00
20		2.3 Electrical	80 days	12/11/2018	4/29/2019	12	0%	500		0 hrs	$0.00
21		3. Project No. 3 Transformer and Sub...	271.2 days	9/25/2018	1/10/2020		0%	500		0 hrs	$0.00
30		4. Project No. 4 Beta Retirement Co...	252 days	6/6/2018	8/20/2019		0%	500		0 hrs	$0.00

Figure 7-23 Baseline Schedule Updated with Project No. 2 Piling Activities

7.1.4 Project Controlling and Monitoring

Site Roads and Site Office Work

On May 18, 2018, the road subcontractor commenced the excavation of the areas required for the site access roads. These would only be gravel roads, but an extensive amount of excavation was required, which had to be done within 10 days so that the roads would be completed within the 30-day window specified in the Retirement Community Beta Project First Milestones Baseline Schedule.

After four days had passed, the subcontractor's Site Manager reported to the Senior Project Manager, Bob, and his team that they had completed removal of 565,000 ft³ of dirt and felt that they would be completed with the excavation within the next six days. Bob was told by the Site Construction Management Team that this contractor was falling behind schedule because, according to the analysis in the project's CEP, this latest actual progress was 26.8%, which was less than the planned 30%. The No. 2 Site Construction Manager in the weekly Project Control Meeting explained the analysis performed in Figure 7-24 and displayed the current progress graphically, which confirmed the current slippage in schedule.

Bob requested that the Construction Management Team conduct an analysis of the subcontractor's excavation work to determine what could be improved to ensure that this work would be done in 10 days, as planned. The four Construction Managers first established possible root causes for this slippage in schedule, which were as follows:

1. Insufficient supply of trucks
2. Bottleneck in the process of loading and unloading of each dump truck
3. Insufficient number of working hours

Item No. 1 Analysis

Excavation Amount per Day to Be Removed (YD³) *	Average Capacity of Dump Truck (YD³)	How Many Loads Can Be Effectively Completed per Truck in One 10-Hour Workday	Number of Trucks Required Onsite per Day for Excavation
7,822	10	12**	= 7822/10 × 12 = 65

*   27 ft³ = 1 YD³
** Assuming it takes 50 minutes for the whole process of loading and unloading of the truck

A check onsite revealed that the subcontractor only had 62 dump trucks onsite at this time but was trying to bring in 5 more trucks to improve the current situation.

Retirement Community Beta Project Project Construction Execution Plan (CEP) - Excavation Worksheet for Roads

Excavation Planned Duration (Days)	10				
	Distance of Roads	5.00 Miles	26,400.00 Feet		
Excavation Amount Required	Depth of Excavation	5 Feet		LxDxW	2,112,000.00	Cubic Feet (FT^3)
	Width of Excavation	16 Feet			
Per Day Required to Meet Schedule	211,200.00	Cubic Feet (FT^3)			

Graph

	Day 1	Day 2	Day 3	Day 4	Day 5	Day 6	Day 7	Day 8	Day 9	Day 10	Total
Planned	211,200	211,200	211,200	211,200	211,200	211,200	211,200	211,200	211,200	0	2,112,000
Actual	150,000	200,000	215,000								565,000

Progress to Date	26.8%
Planned To Date	30.0%

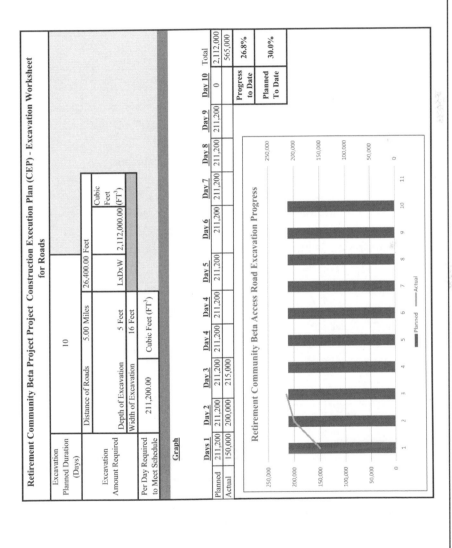

Retirement Community Beta Access Road Excavation Progress

Figure 7-24   Site Access Road Construction Worksheet

Item No. 2 Analysis

Activity	Duration (Minutes)	Comments
Truck Waiting to Be Loaded	10	Usually four to five in the queue.
Loading of the Truck	30	Using two Excavators and 10 minutes to back into position.
Truck Travels to Spoils Dump Location	15	This is one way from the construction site.
Truck Unloads	7	Small amount of positioning.
Truck Returns to Construction Site	12	
Total	74	This is for current configured path.

This reduces the loads per a 10-hour work day to 8 loads per day. This would change the number of trucks required to 97, which is beyond the capacity of the subcontractor. The four Construction Managers presented this information to the subcontractor's Site Manager, and together they changed the loading of the trucks to a two-line approach from the current one-line approach, which reduced the loading time from 30 minutes to just 15 minutes.

Activity	New Durations (Minutes)	Comments
Truck Waiting to Be Loaded	10	Usually four to five in the queue.
Loading of the Truck	15	Using four Excavators and 5 minutes to back into position.
Truck Travels to Spoils Dump location	15	This is one way from the construction site.
Truck Unloads	7	Small amount of positioning.
Truck Returns to Construction Site	12	
Total	59	This is for current configured path.

This revised loading process increased the loads per a 10-hour work day to 10 loads per day, which would increase the number of trucks required to 78. The subcontractor's Site Manager stated that the company will provide the additional 16 trucks within two days and would also take action to reduce the truck's waiting time to five or seven minutes, which would further help the current situation.

Item No. 3 Analysis

If the subcontractor would work an additional two hours per day, which would not require any night-time operation, the gain in unloading with 78 trucks in the current configuration would be as follows:

Hours per Day	Number of Trucks	Average Capacity of Dump Truck (YD³ × Loads)	Amount of Material Removed per Day
10	78	$10 \times 10 = 100$	7,800
12	78	$12 \times 12 = 144$	11,232

The subcontractor's Site Manager informed the EPC Construction Managers that their current project budget would not be able to cover the additional cost of two hours of overtime for each of the 78 drivers because the additional cost of 13 trucks and 2 excavators per day was already being covered by the project's contingency funds.

The Construction Management Team presented the results of their analysis to the Senior Project Manager, Bob, along with the feedback from the subcontractor's Site Manager, and the decision was made that a verification of the number of trucks, excavators, and the current time for each process would be made at least once per day by one of the Construction Managers until the excavation work was completed.

Site Offices

On June 1, 2018, access to the area for construction was completed—one day ahead of schedule—and the customer's approval of the office-layout drawing was complete, as shown in Figure 7-25.

In this figure, it can be seen that the Document Control Center is separated from the main offices. This is required due to its high level of noise, which can be disruptive in an office, and the much higher HVAC requirements. The Senior Project Manager, Bob, ensured that the Site QA/QC Manager, Robert Lieberstein, was directly involved with its design because it is one of the most critical areas on the project: The loss of one QA/QC record can result in the company being unable to close out the project. Bob had memories of one project where the company could not obtain the necessary Certificate of Occupancy for a seven-story apartment building because the QA/QC Department had lost the signed-off Request for Inspection (RFI) on the apartment's firefighting system

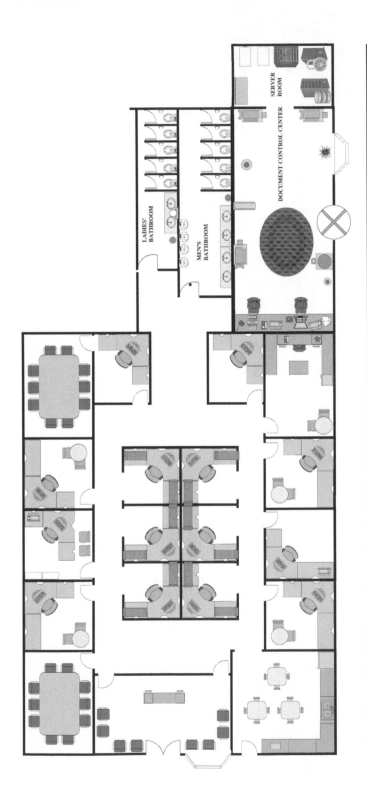

Figure 7-25 Layout Drawing for the Site Offices

when the subcontractor performed the final leakage test, which prevented closure of the project until the test was performed again. This leakage test was a sunk cost for the project, and it was a very high cost for the company because it also affected its reputation in the construction industry for a few years. After the excavation was completed and the backfilling along with compaction began for the site office slab foundation, the No. 4 Site Project Manager, Debra Wolfe, focused on ensuring that the building services such as sewage, potable water, electrical service (portable generator), communication, and internet via fiber optics were on track to be completed at the time the permanent site office would be completed. The other area she focused on was working the company's Procurement Office to coordinate all of the office equipment and furniture so that its delivery would ensure that the EPC Company's Site Team had a completed office to move into on July 4, 2018. In addition to this new permanent site office, the Senior Project Manager, Bob, had four portable office trailers delivered to the site so that each project had one mobile project office at the construction site for both the site engineers and management because subcontractor surveillance is critical to completing this project on time and under budget.

Solar-Generating Facility (Project No. 1) and Wind-Generating Facility (Project No. 2)

The Civil Work for Project Nos. 1 and 2 started on June 6, 2018, because the Project Management team was concerned that the additional piling work could possibly impact the project's schedule. The impact comes from the fact that both projects must be electrically completed at the same time in order for Project No. 3, which is the transformer building, to be connected to the existing 300-KV electrical grid by the local utility. The schedule for the Civil Work on both projects is shown in Figure 7-26.

This Project Level 5 Baseline Schedule, which was produced by the Project Control team, is a Project Management System deliverable, but it comprises inputs from both the Work Management and Quality Management Systems. The Solar-Generating Facility Construction Manager, Clark Montage, and Eric Floutaff, the Wind-Generating Facility Construction Manager, explained this to their project teams and why the failure to effectively manage the work activities along with their respective quality sign-offs can severely impact not only the progress of both projects but also the whole project because all of the projects are linked together in this Level 5 Baseline Schedule. This linkage is evident by the arrows that link specific activities in one project with the start of another activity in another project. For example, as Clark stated to the group, it is evident that the completion of the Benchmark activity is required before the first Civil Activity can start for Project No. 4, when one follows the arrow from

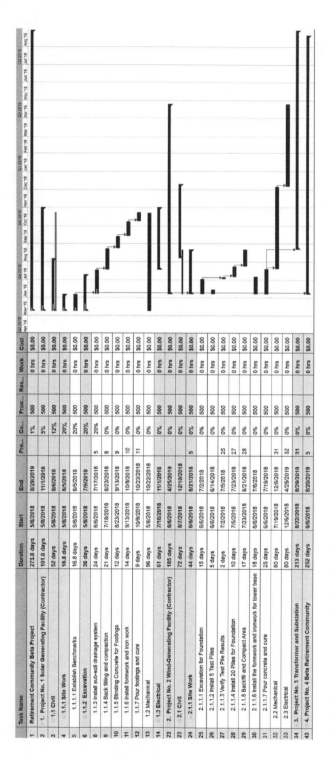

#	Task Name	Duration	Start	End	Pre...	Co...	Prior...	Res...	Work	Cost
1	Retirement Community Beta Project	273.8 days	5/8/2018	8/29/2019		1%	500		0 hrs	$0.00
2	1. Project No. 1 Solar Generating Facility (Contractor)	101.8 days	5/8/2018	11/1/2018		5%	500		0 hrs	$0.00
3	1.1 Civil	52 days	5/8/2018	8/6/2018		12%	500		0 hrs	$0.00
4	1.1.1 Site Work	16.8 days	5/8/2018	6/5/2018		20%	500		0 hrs	$0.00
5	1.1.1.1 Establish Benchmarks	16.8 days	5/8/2018	6/5/2018		20%	500		0 hrs	$0.00
6	1.1.2 Excavation	36 days	5/8/2018	7/9/2018		20%	500		0 hrs	$0.00
8	1.1.3 Install sub-soil drainage system	24 days	6/6/2018	7/17/2018	5	20%	500		0 hrs	$0.00
9	1.1.4 Back filling and compaction	21 days	7/18/2018	8/23/2018	8	0%	500		0 hrs	$0.00
10	1.1.5 Blinding Concrete for Footings	12 days	8/23/2018	9/13/2018	9	0%	500		0 hrs	$0.00
11	1.1.6 Install formwork and iron work	14 days	9/13/2018	10/8/2018	10	0%	500		0 hrs	$0.00
12	1.1.7 Pour footings and cure	9 days	10/8/2018	10/23/2018	11	0%	500		0 hrs	$0.00
13	1.2 Mechanical	96 days	5/8/2018	10/22/2018		0%	500		0 hrs	$0.00
14	1.3 Electrical	61 days	7/18/2018	11/1/2018		0%	500		0 hrs	$0.00
22	2. Project No. 2 Wind-Generating Facility (Contractor)	185 days	6/6/2018	4/25/2019		0%	500		0 hrs	$0.00
23	2.1 Civil	72 days	8/7/2018	12/10/2018		0%	500		0 hrs	$0.00
24	2.1.1 Site Work	44 days	6/6/2018	8/21/2018	5	0%	500		0 hrs	$0.00
25	2.1.1.1 Excavation for Foundation	15 days	6/6/2018	7/2/2018		0%	500		0 hrs	$0.00
26	2.1.1.2 Install 5 Test Piles	5 days	6/6/2018	6/14/2018		0%	500		0 hrs	$0.00
27	2.1.1.3 Verify Test Pile Results	2 days	7/2/2018	7/5/2018	25	0%	500		0 hrs	$0.00
28	2.1.1.4 Install 20 Piles for Foundation	10 days	7/5/2018	7/23/2018	27	0%	500		0 hrs	$0.00
29	2.1.1.5 Backfill and Compact Area	17 days	7/23/2018	8/21/2018	28	0%	500		0 hrs	$0.00
30	2.1.1.6 Install the formwork and ironwork for tower base	18 days	6/6/2018	7/6/2018		0%	500		0 hrs	$0.00
31	2.1.1.7 Pour concrete and cure	25 days	6/6/2018	7/19/2018	31	0%	500		0 hrs	$0.00
32	2.2 Mechanical	80 days	7/19/2018	12/6/2018	31	0%	500		0 hrs	$0.00
33	2.3 Electrical	80 days	12/6/2018	4/25/2019	32	0%	500		0 hrs	$0.00
34	3. Project No. 3 Transformer and Substation	213 days	8/22/2018	8/29/2019	31	0%	500		0 hrs	$0.00
43	4. Project No. 4 Beta Retirement Community	252 days	6/6/2018	8/20/2019	5	0%	500		0 hrs	$0.00

Figure 7-26 Retirement Community Beta Project No. 1 and No. 2 Civil Activities

Activity No. 5 to Activity No. 43. After this meeting was completed, the two Construction Managers set up their Site Construction Teams on a two 8-hour shift basis so that the Civil Work now in progress onsite could be monitored closely for the next two months until all of it was completed because all of the civil subcontractors were working 10- to 12-hour shifts per day on a 5-day-per-week basis.

Retirement Community Beta (Project No. 4), Solar-Generating Facility (Project No. 1), and Wind-Generating Facility (Project No. 2) Lifting Plans

The Senior Project Manager, Bob, was informed by the Site QA/QC Manager, Robert Lieberstein, that the customer signed off the ITP for the Solar Panels and the associated structural components last Friday in the factory, which should permit all of the associated vendors to ship these components to the site sometime in July. Robert also stated that the customer's Engineer finished the inspection and testing process with the manufacturer on all of the Wind Turbine components, including the generator, and this ITP is now being signed off, which will allow these components to be shipped in August. After this information was received, Bob, asked each Project Manager for Project Nos. 1, 2, and 4 if they had received a Lifting Plan for all of the equipment coming to the site, which would include unloading the components and the erection of the structures for the equipment. Debra Wolfe, the Project Manager for the Retirement Community Beta, stated that she and her team had the Lifting Plan from the contractor, which was the same one shown in the project's CEP because the same subcontractor was involved in the construction of the building. She explained that the Lifting Plan, which was also incorporated in the project's Health, Safety and Environmental Plan (HSE Plan), involved the erection of two tower cranes on both sides of the construction site. Debra also stated that their foundations, along with the electrical duct work, was currently in progress because the customer signed off on both the Lifting Plan and HSE Plan for Project No. 4 last month. However, the Project Managers for Project Nos. 1 and 2 had not received any type of Lifting Plan or a related HSE Plan from either the Solar- or the Wind-Generating contractors.

The Senior Project Manager, Bob, after receiving this feedback, set up a one-hour meeting with the Wind-Generating subcontractor's Project Manager and the Solar Generator subcontractor's Project Manager. This meeting also included the Site Project Management team and the Site HSE Manager, Sandra Chonelly.

At this meeting, the Solar-Generating Facility Project Manager provided a copy of their Lifting Plan along with their HSE Plan to the Senior Project

Manager and staff, but stated that their lifting was primarily just for the Solar Panel frames after the support structure is completed, which is performed with a state-certified 20-ton mobile crane. The Solar Panels are offloaded with a special fork truck, when they arrive, and are then set in place by hand, after being unpacked. The 4-ton Inverter/Transformer will arrive in a separate shipment and will be set in place with this crane, once it arrives. She added that in the HSE Plan all of the safety risks associated with this work are clearly identified, and each one has a mitigation plan, which is then discussed with the workers by the foreman before the respective work begins. She also confirmed that the crane driver or any operator involved with lifting equipment is a certified operator for the equipment and has a valid Arizona license. The last item she added was that all of the workers, including the supervising foremen, are certified riggers, and their training records are on file in the company's site office. Sandra requested that the Solar-Generating Facility Project Manager provide a copy of these operating licenses along with the fork truck and crane's certification for the company's site HSE records.

The Wind-Generating facility Project Manager, after handing the Senior Project Manager his company's Lifting Plan and HSE Plan, gave the following presentation to all of the meeting attendees:

1. **Slide One** – This will be a 1-MW Wind Turbine Generator that we will install. Just the supporting structure for this generator is approximately 240-feet high. At the very top of this structure, we will first install a 20-ton generator housing that includes the gear box, which will be driven by a series of fan blades that are connected to a hub mounted on the shaft. These fan blades are first assembled on the hub, and then this whole fan assembly is installed on the shaft after the generator is lifted and installed on the support structure.

2. **Slide Two** – Our company employs a state-certified rigging company just for the offloading of all components and their subsequent erection. This lifting equipment will be a 60-ton portable crane for the structure base and the three subsequent sections that are set and bolted in position. This crane will also be used to offload the fan blades and hub, which will be followed by their assembly. The heavy lifting will be performed by a special 600-ton crawler crane that will be positioned so that the three support sections, the generator house, and fan assembly can be quickly installed in sequence. This large crane will be delivered to the site in a series of sections along with the crawler crane and take three weeks to assemble. I will provide a copy of the state certification for both of these cranes along with the certifications for this rigging company and its crane operators to Sandy at the end of this meeting.

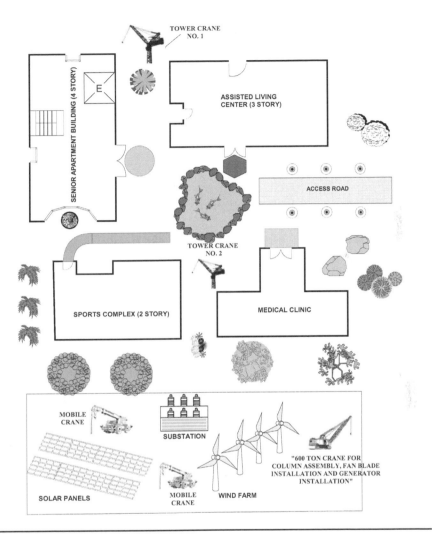

Figure 7-27   Retirement Community Beta Construction Site Lifting Plan

(See text describing Figure 7-27 on the following page.)

3. **Slide Three** – All of our workers involved are certified riggers with previous experience on similar projects, in many states, including Arizona. Their certifications are on file in our site office. All of the rigging equipment is also certified, and those certificates will also be provided to Sandy at the conclusion of this meeting. On the day of the large 20-ton lift, we will have a Tool Box Safety Briefing for all onsite personnel involved with the heavy lift, and this will include an attendance sheet that must be signed by all participants. If anyone is not a participant, they will be

asked to stand behind the safety barricades at the designated areas and not to interfere with the lift, once it starts.

4. **Slide Four** – Does anyone have any questions on what I just presented; if not, thank you for this meeting and please contact me immediately if there any questions regarding our Lifting Plan or HSE Plan.

The Senior Project Manager, Bob, concluded the meeting after the Site HSE Manager, Sandra Chonelly, confirmed that everyone signed the meeting attendance roster, and she had received all of the documents discussed, from both of the subcontractor Project Managers. The Wind-Generating Facility Project Manager, Lawrence Valenko, compiled the MOM, which included a Site Lifting Plan Sketch as shown in Figure 7-27, and had all of the required Project Managers, including the Senior Project Manager, review and sign it prior to its submittal to both subcontractor Project Managers for their final acknowledgment and approval the next day.

7.2 Concluding Remarks

We will leave Bob and his Site Project Team now as they continue to build and finish the Retirement Community Beta Project in Arizona and move onto the next chapter. In this chapter, the author will share many very valuable "lessons learned" from thirty years of experience in the power and construction industries, which involved working on remote constructions sites around the world with international project teams.

Chapter 8

Lessons Learned from the Field

8.1 Introduction

In the construction industry there are a number of Project Management System, Work Management System, and Quality Management System tools that are not being employed on various projects worldwide, which is having a negative effect on many projects from the standpoint of cost and schedule. It is the intent of this chapter, with a couple of lessons learned for each of these three systems, to show how some of these tools can have a very strong, positive impact on a project—an impact that will also greatly benefit the Project or Program Manager who is directly responsible for the project or group of projects.

8.2 Project Management Systems Lessons Learned

A. Baseline Schedule

On one very large, complex construction project in a foreign country, the author was made aware by the Owner of the project, which he represented as the Site Manager, that the EPC Company still had not signed the Level-Two Baseline Schedule for the project. This was despite the fact that construction work had begun almost one year ago and the initial Baseline Schedule was presented at the beginning of the project. An examination of the Level-Three

Contract Schedule for the monthly progress reports, which were provided to the Owner by the EPC Company, revealed that the number of Critical Path activities were increasing beyond those presented in the initial Level-Two Baseline Schedule. A discussion of how this can happen and what it represents is presented in Figure 8-1.

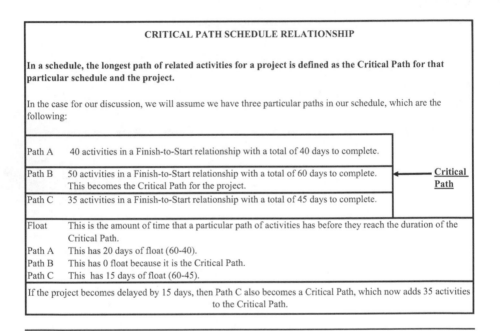

CRITICAL PATH SCHEDULE RELATIONSHIP

In a schedule, the longest path of related activities for a project is defined as the Critical Path for that particular schedule and the project.

In the case for our discussion, we will assume we have three particular paths in our schedule, which are the following:

Path A　40 activities in a Finish-to-Start relationship with a total of 40 days to complete.

Path B　50 activities in a Finish-to-Start relationship with a total of 60 days to complete. ◄──── **Critical Path**
This becomes the Critical Path for the project.

Path C　35 activities in a Finish-to-Start relationship with a total of 45 days to complete.

Float　This is the amount of time that a particular path of activities has before they reach the duration of the Critical Path.
Path A　This has 20 days of float (60-40).
Path B　This has 0 float because it is the Critical Path.
Path C　This has 15 days of float (60-45).

If the project becomes delayed by 15 days, then Path C also becomes a Critical Path, which now adds 35 activities to the Critical Path.

Figure 8-1　Discussion of Critical Path Schedule Relationship

A negotiation process began between the Owner and the EPC Company to resolve this issue and finalize a Level-Two Baseline Schedule that both parties finally signed. After one more year passed, with the project falling further behind and limited success being achieved on the completion of numerous activities, the EPC Company filed a large financial claim against the Owner for delaying their initial site access as per the newly signed Baseline Schedule. However, the analysis that was performed by our Project Control team onsite revealed that extensive soil improvement work that was required before any civil foundation work could proceed was the culprit, not the small delay in the site access. A further investigation revealed that EPC Company had failed to perform a proper onshore soil analysis at the time of Project Planning that would have identified this major problem early in the project. This finding, along with the schedule analysis, resulted in this claim being rejected by the Owner.

Lesson Learned

As the Project or Program Manager, the first order of business must be the establishment of a project Baseline Schedule, which must be signed by both the Owner or customer and the respective EPC Company. This must be done during the Project Initiation phase and before any construction activities are begun for the following reasons:

- It provides a solid foundation for the Project Management System outputs presented earlier in this book.
- It also provides a solid foundation for the Work Management System and the Project Quality System in the Project Planning phase, which will improve their effectiveness in the Project Execution phase when the construction work starts onsite.
- It ensures accountability on the part of both the Owner or customer and the EPC Company to ensure that the milestone dates, as set forth in the Baseline Schedule, are maintained over the life of the project.
- It permits alignment to commence for all of the project's deliverables at the time of the Project Initiation phase so that the processes necessary to maintain this alignment can be developed in the Project Planning phase, which will improve their effectiveness and communication between all project stakeholders once the work begins on a large, complex construction project.

B. Communication Plan and Transparency

After the Project or Program Manager identifies all of the stakeholders, along with their communication preferences, a Project Communication Plan can be developed and compiled that both the customer or Owner and the EPC Company must approve upon review. This can take a number of iterations until all parties give final approval and must be done in the Planning Phase of the project. However, this Communication Plan becomes just a piece of paper unless there is transparency in the relationship between the Owner or customer and the EPC Company on a large, complex construction project.

On one project in which this author was the Owner's Construction Manager, the Owner spent countless sums of money to promote teamwork between their team and the EPC Company's onsite management team. However, the EPC Company's site management did not feel that they had to be transparent regarding their performance onsite, despite these team-building workshops. For example, a subcontractor working for the EPC Company was found by the Owner's Site Engineers performing work without the required "Issued for Construction" drawings. The EPC Company, when confronted with this finding, stated that the drawings in questions were almost completed, but they decided to proceed with the work since it was already behind schedule. The Owner was not happy with this finding, but was more disturbed that the EPC Company did not approach them immediately to discuss this problem so that both parties could reach a "win-win" solution that would benefit everyone and the project. In another instance on this same project, the EPC Company ran out of a critical chemical that was required for the fabrication of a large number of wafer-type subassemblies, but remained silent and tried to procure this chemical on their own. However, they failed continually to get the local Custom's clearance for this chemical so that it could be brought into the country and to the construction site. This problem was only found when the Owner's Construction Manager requested an explanation for the 30- to 40-day delay on the activity, requiring this material from the EPC Company's Site Manager. After this problem was reported by the EPC Company's Site Manager, the Owner was able to resolve it in one week with the country's Custom Office. It is incidents like these, when the EPC Company feels that they can work without being transparent, that the project, along with the site relationship between the Owner or customer and the EPC Company, suffers as the trust between the two parties continues to break down.

Lesson Learned

The Project Communication Plan must be developed at the time of Project Initiation because many customers or Owners want to see this at the time of bid submittal, which will be developed further in the Planning Phase for the project. The critical items that this Project Communication Plan must cover are the following:

- It must promote the concept of "One Team–One Goal" for the EPC Company and their subcontractors, but at the same time do the same for the Owner's or customer's onsite Project team.
- It must include a one-hour site tour by the EPC Company's Construction Manager and the Owner's or customer's Construction Manager to see firsthand the progress onsite and spot any issues early, before they get worse. For example, if the scaffolding subcontractor is stating in the Site Weekly Progress meeting that they are 60% complete, but this tour reveals they are only 40% complete. The two Construction Managers can go directly to this subcontractor and have their Site Manager explain this delay, along with a plan to recover the lost time.
- It must have provisions in it that will involve team-building events over the life of the project, which should include both the customer's or Owner's and the EPC Company's Site Management team.

- It must ensure that all site meetings are limited to one hour and that all action items in the Minutes of Meeting (MOM) have a date for completion along with the name of the person responsible.
- It must establish a document-tracking system for all documents onsite to ensure that alignment is maintained throughout the project at the site and in the corporate office of both the Owner or customer and the EPC Company.

8.3 Work Management Systems Lessons Learned

A. Resource Allocation

In the construction industry today, one of the major issues that destroys a project's Baseline Schedule and its budget is the failure to plan and accurately allocate the resources for a project. In many meetings, as a Site Construction Manager or Site Manager for the Owner, this author sees a resource allocation chart by the EPC Company or its subcontractor, which is similar to the example shown in Figure 8-2.

An examination of the chart in Figure 8-2 reveals that this Civil Subcontractor is consistently failing to allocate resources to meet their planned monthly resource levels, and they are consistently falling behind. This not only a problem, but it will force this Civil Subcontractor to now shift from doing the scheduled work to doing "what they can work." On one project, this monthly deficiency comprised thousands of workers, and the net result was a project that led to the near bankruptcy of the EPC Company and an 18-month delay in the project.

This project and many others clearly show that resource allocation is not being performed at the time of Project Initiation, but if it was, then the Project or Program Manager should see something similar to what is shown in Figure 8-3, which is an example of just two civil activities that have been resource loaded along with their costs.

If an EPC Company has a well-organized and written Construction Execution Plan (CEP) for the project, then building a Baseline Schedule with the information shown in Figure 8-3 for each activity is not difficult.

Lesson Learned

If there is no accurate resource allocation performed at the time of Project Planning, then not only does the Work Management System fail because subcontractors are no longer completing work as scheduled with the right amount of resources and equipment. The Project Management System

XYZD CIVIL SUBCONTRACTOR RESOURCE CHART

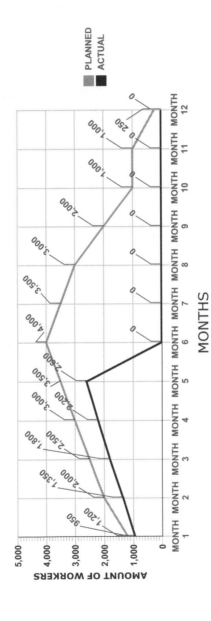

	MONTH 1	MONTH 2	MONTH 3	MONTH 4	MONTH 5	MONTH 6	MONTH 7	MONTH 8	MONTH 9	MONTH 10	MONTH 11	MONTH 12
PLANNED	1,200	2,000	2,500	3,000	3,500	4,000	3,500	3,000	2,000	1,000	1,000	250
ACTUAL	950	1,350	1,800	2,200	2,600	0	0	0	0	0	0	0

Figure 8-2 Example of Civil Subcontractor Resource Loading Chart

Example of Two Civil Construction Activities Resource Loaded with Cost

Branch Name	Duration	Start	End	Priority	Completion	Resources	Work	Cost
1. Civil Foundation Work	125 days	2/14/2018	6/26/2018	500	1%		1060 hrs	$41,800.00
1.1 Bending Rebar	50 days	2/14/2018	4/10/2018	500	1%	General Labor[75%];Bar Benders[40%]	460 hrs	$15,400.00
1.2 Erect Formwork	75 days	4/10/2018	6/26/2018	500	1%	Carpenters[40%];General Labor[60%]	600 hrs	$26,400.00

Figure 8-3   Example of Two Civil Construction Activities Resource Loaded

also fails because the durations of all Critical Path activities, as originally scheduled, are much longer than scheduled, and the project's budget man hour expenditures increase beyond the established limit. To prevent this from happening, the following actions must be taken by both the Project or Program Manager and the Construction Manager:

- Once the workscope is identified in the Project Initiation phase, spend the time and money to develop a well "thought-out" and very accurate Construction Execution Plan (CEP) for the project. After this is done, then update it continually over the life of the project so that the company will have historical data to use for future projects.
- Ensure that each subcontractor brings not only the right amount of resources and equipment for the job but also enough site foremen or front-line supervisors so that there is one supervisor for every seven people. This will prevent a lot of people onsite from "standing around" because they do not know what to do next.
- If you are working in a foreign country that requires a specific percentage of local workers to be in your resource group, start the hiring process early by requesting the Owner or customer to provide a list of reputable labor-supply agencies in the local area.
- If you are having trouble obtaining enough union workers in an area of the United States, contact the respective union business agents and request their support in increasing the allocation of union workers that can be brought in from outside the local union, who are called "travelers."
- When it comes to resources, be proactive and realize that the site permitting and safety training for each worker has to be done before they start work, which in some cases can be up to two weeks. This cost should appear in the project's budget as "resource mobilization" expenditures.
- Examine the site and establish the work break areas along with restrooms prior to the workers arriving, and locate these portable facilities to support the work onsite.
- Spend the time to make sure that every worker has all of the proper safety training for the work they will be performing, which will enhance the project's overall site safety performance.
- After the project Baseline Schedule is completed, make sure that the Site Engineers and Construction team are in agreement with the durations and resources shown for each activity.

B. Movement of Materials and Equipment

On one project, as the Owner's Construction Manager, this author watched a very large foundation being erected with thousands of rebar steel pieces of various shapes and sizes, which were manually brought from the bending yard to the foundation for installation. If we examine this process, there are two "choke" points—the bending machine and then the movement of each bended piece of rebar to the foundation. The subcontractor increased the bending machines to two, thereby reducing the impact of the first "choke point." But the rebar pieces were still being manually moved to the foundation using a small group of general laborers. In a discussion with this subcontractor's Project Manager, this author suggested that he give thought to bringing in a small mobile crane along with a flatbed truck so that the increased production from both bending machines could be easily and quickly transported directly to the laborers installing these pieces in the foundation. The Project Manager followed this advice, and the large foundation was erected 10 days earlier than planned.

Lesson Learned

- Make a plan to walk the full construction site each day, spending a few hours or more examining how various work activities are being performed. After returning to the office, evaluate those activities that will have the greatest impact on the project's Critical Path activities to see if any improvements can be made to finish them more quickly.
- The Construction Manager, and the Site Engineering team, along with the Project or Program Manager, should evaluate the materials and equipment that will require installation onsite and establish a Site Lifting Plan in the project's Planning Phase. If the construction site has limited access in many areas, consideration should be given to the strategic placement of various tower cranes with the correct lifting capacity required for the materials that will be installed.
- Evaluate each subcontractor that comes to the site and locate their site office as close as possible to the area where they will be working.
- Always use the philosophy of "bringing the materials or equipment to the workers at their location," instead of bringing the same worker to the material and then have them return with this material to their workplace.
- Effectively plan all heavy lifts and physically walk down the path or road that will be utilized to ensure that it is ready for this lift when scheduled. If areas need to be strengthened, this may take an additional two weeks of work, so always keep this in mind when evaluating dates for all heavy lifts on a large construction site. An "engineered lift" are those above 50 tons and usually require a qualified heavy lift subcontractor with certified rigging personnel and equipment.
- Make sure that all lifting equipment brought to the site has the standard certifications and that all rigging personnel have had the required safety training for all lifting activities. This certification includes the lifting

slings, which must have current load-testing certificates that must be kept in the EPC Company's HSE files onsite.

8.4 Quality Management Systems Lessons Learned

A. Request for Inspection (RFI)

In the construction industry there is always confusion over Quality Assurance and Quality Control, which can easily be explained in the following two documents:

- **Method Statement (MS)** – This document is the Quality Assurance procedure, which is comprised of an aggregate collection of all of the activities, defined in steps that must be performed in the correct logical order and in accordance with all related specifications to produce the quality of the deliverable as defined in the contract.
- **Request for Inspection (RFI)** – These Quality Control documents, which can be quite a few for just one method statement, are the series of tests and inspections, which are laid out in steps. The inspections in these RFIs must be performed and approved by the EPC Company, the Owner, or both before the next activity in the Method Statement can be performed.

During the execution of construction activities onsite, all of the Method Statements, which have been reviewed and approved by either the Owner or the customer and the EPC Company, must be followed by each subcontractor. The completion status of each Method Statement can be easily determined by dividing the number of actual RFIs that have been completed by the total amount of planned RFIs.

This author has been on various construction projects wherein an RFI Completion Report from the EPC Company, which is produced from a subcontractor's feedback, is shown that is similar to that presented in Figure 8-4.

The greatest concern during construction is the number of RFIs that have been rejected, which is done by the Owner's or customer's Site Engineer, for the following reasons:

a. The work was not completed.
b. The work did not meet quality specifications.
c. The subcontractor was not present at the time of the inspection.
d. The area for the inspection was unsafe.

The first item in the list above is a great concern to the EPC Company because it will entail rework by the subcontractor, which is a cost that they must absorb and could impact the scheduled completion of the foundation.

Branch Name	Duration	Start	End	Priority	Completion	Resources	Work	Cost
1. Civil Foundation Work	125 days	2/14/2018	6/26/2018	500	1%		1060 hrs	$41,800.00
1.1 Bending Rebar	50 days	2/14/2018	4/10/2018	500	1%	General Labor[75%];Bar Benders[40%]	460 hrs	$15,400.00
1.2 Erect Formwork	75 days	4/10/2018	6/26/2018	500	1%	Carpenters[40%];General Labor[60%]	600 hrs	$26,400.00

RFI Status for the Erection of Foundation ABZ (WBS 1.0)

Column1	Column2	Column3	Column4	Column5
Week No.	Planned	Done	Rejected	Not Done
1	50	30	10	10
2	65	55	12	8
3	70	77	5	
4	75	73	4	3
5	65			

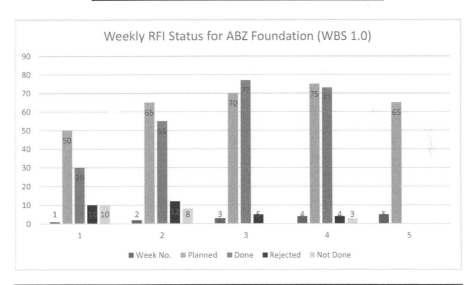

Figure 8-4   Example of RFI Completion for a Large Foundation

The second item can lead to a Nonconformance Report (NCR) being filed by the Owner or customer, which may require the EPC's Engineers to work with the QA/QC Manager to resolve and can damage the reputation of the EPC Company. The third item may suggest that the subcontractor does not have enough personnel or management onsite, which the EPC Company should investigate and have resolved quickly. This last item is serious because the EPC Company is always working hard to ensure there will be no accidents onsite, and this could suggest that the subcontractor is not in compliance with the Site Safety Plan.

Lesson Learned

If the number of RFIs being performed starts to accumulate a large backlog of rejected RFIs or RFIs not being completed, the EPC Company must immediately investigate to determine the cause or causes for this backlog. This backlog will impact both the Project Management System and the Work Management System with a delayed schedule, increased sunk costs, and a greater demand for resources and equipment to start clearing it up. Some of the steps that can be performed early to mitigate this onsite are the following:

- The Owner's or customer's onsite Project or Program Manager must make it very clear to the EPC Company's Project Manager that they are responsible to ensure that all of the required work is done before an RFI is issued to the Owner or customer.
- The EPC Company must audit every subcontractor prior to the start of their work onsite to ensure that they have the QA/QC and HSE staff to ensure compliance with the site's QA/QC and HSE procedures.
- The EPC Company's QA/QC Manager should perform unannounced Quality Audits on various subcontractors during the time of an inspection to ensure that the work is ready for the Owner's or customer's Site Engineer to inspect.
- The EPC Company's Site Management team need to take immediate action when a number of rejected or incomplete RFIs is reported, so that the underlying causes can be identified and corrected.

B. Approved for Construction (AFC) Drawings

During the Engineering Phase of the project, when the Design Engineering firm for the EPC Company starts to develop the layout drawings, followed by the P&IDs, Electrical One-Line Diagrams, Isometrics, etc., each one has to be presented to the Owner's or customer's Engineering team for review and acceptance. The expected output of this intense engineering effort is the production of Approved for Construction (AFC) drawings or Issued for Construction (IFC) drawings, as some companies like to call them. These are also QC documents that are used not only to effectively manage excavations and erect various structures, which includes installing related equipment, but also to ensure that the final product meets all of the required contractual quality and engineering standards. This author has been involved with projects in which the EPC Company has tried to revise various components or structures without resubmittal of the respective AFC or IFC drawing to the Owner's or customer's Engineering team for review of this proposed design change. The primary motive is usually cost driven due to the cost of revising drawings, which is also an internal sunk cost for the EPC Company. The other issue is the failure of a document-tracking system to ensure that only the correct revision of the latest AFC drawing is being used by all subcontractors for the work they are performing onsite. On one project, the construction of a large structure was performed without the latest revision of the respective AFC drawing, and when the RFI using a later drawing revision for the support column installation was performed, the support columns were now found to be in the wrong location. This required a major review of the situation by the EPC Company's Design Engineering firm to resolve the problem and provide an MS for the relocation of these columns. The impact to the overall complex construction project was a three-month delay, and the EPC Company was forced to pay for all of the associated costs.

Lesson Learned

Step 1: This quality problem can only be resolved by aligning the Document Control System that is used by both the Owner or customer and the EPC Company, which can be accomplished by incorporating the WBS Number for each activity in the project's Baseline Schedule with its respective drawings and quality procedures. This step, along with the other steps below, is crucial in ensuring that all work onsite is being performed in accordance with the correct drawings, specifications, and procedures.

Step 2: Establish a site-based Document Control Center for the both the Owner or customer and the EPC Company.

Step 3: Ensure that both parties have a common database and a Master Drawing List (MDL) that is continually updated, which should also contain a respective WBS Number that relates to the drawing.

Step 4: The QA/QC Manager from both groups should have a designated person or persons on their respective team that monitors all RFIs to ensure that the drawing and its revision are correct prior to its issuance. This can only be done by placing the MDL on a secure designated server, which is located onsite and which both Document Control Centers can access for updates and retrievals.

Step 5: The EPC Company's Site Engineering Team must also audit the MDL to ensure that all of the required vendor drawings are available for equipment and material receipt inspection by the QA/QC Department.

Step 6: The EPC Company and the Owner or customer QA/QC Manager should conduct random audits of the subcontractor's facilities to ensure that they are using the correct AFC drawings and that they remove those drawings that are outdated.

8.5 Closing Remarks

This last chapter concludes this book, which has been written with the hope that its readers, upon completion of reading, will enjoy and feel confident about building the future one step at a time with the tools provided in this book, which is exactly what most large, complex construction projects are all about.

Glossary

AFC (Approved for Construction) – This is the status of a drawing that permits it to be used for construction purposes onsite. It is a drawing that has been reviewed and approved without comment by both the EPC Company's Engineering team and the customer's or Owner's Engineering team.

BOM (Bill of Material) – After a drawing is AFC, this document is the list of the material that will be required to complete the system or equipment installation shown on this drawing.

CCB (Change Control Board) – This is a board within the EPC Company or a customer's or Owner's company that will review any project contract Change Order Request that meets the internal criteria for submittal to the CCB. It usually involves a large amount of money or large impact to the project's Baseline Schedule, but in some cases, it is both. If the CCB rejects the project contract Change Order that the EPC Company has submitted, then the senior management of the customer or Owner's company and the EPC Company will have to negotiate the amount or time being requested.

CEP (Construction Execution Plan) – This is a plan that takes each work activity defined in the project's contractual work scope and breaks it down into the resources, material, and equipment required to complete that specific work activity (explained in greater detail within this book). In most cases, it is good practice to also provide the time and cost to complete that activity, which will make developing the project's budget much easier.

CPM (Critical Path Method) – This is a method of scheduling by taking the activity with the longest duration and using this as a critical path. This means that this activity has no "float" and all of its subactivities must be started and completed as scheduled. All of the other activities and their total duration are then matched to this timeline.

DCN (Design Change Notification) – This is usually a technical request from the EPC Company to the Owner or customer requesting that a particular component, equipment, piping, etc., have its design changed for various reasons. This DCN will require that the drawing or document affected be changed and run through the review process before it can be labeled as AFC.

ECC (Erection Completion Certificate) – This is the document produced by the EPC Company after all of the work, testing, and assembly has been completed for an activity, which includes any "red lined" drawings for small changes of piping, electrical wiring, plumbing, etc., that were performed during the erection of this one system. The ECC is usually a package of many "work packages," all related QC documents, and the "red lined" drawings, previously mentioned. The EPC Company must present each ECC to the Owner or customer and have their approval before the system is declared complete for operation or the building is ready for occupation, in the case of Civil ECCs.

Electrical One-Line Diagram – This is an electrical drawing that is a graphical representation of a three-phase power system in which one line represents all three phases, and specific symbols for the transformer, breaker, grounding, etc., are added, which makes it a very simple block diagram for quick analysis of a specific electrical system because it also contains specific voltage and amperage values at these various points.

EPC (Engineering, Procurement, and Construction) – This is a company performing the construction work onsite that does everything—the Engineering and the Procurement of all materials and equipment, along with all of the Construction Work, which allows an Owner or customer to only have to deal with just one EPC Company. In the past, a large, complex construction project was built by one Construction company and one Engineering Company, called an Architect Engineering firm.

FCRs (Field Change Requests) – On some projects, these documents are also called "Field Change Notifications (FCNs)" or "Field Action Requests (FARs)," and they are submitted by the EPC site quality or construction team when something in the field has to be modified for proper installation and requires immediate approval onsite, so that work is not delayed. The critical item for this document is that the modification to the existing AFC drawing must not change the engineered design of the item or system being modified. All of these

FCRs must be included in the ECC package for each item modified during construction at the time of turnover to the Owner or customer.

FDR (Field Design Request) – This is also called a Field Design Notification (FDN) and is primarily used onsite by the EPC Company's engineers to request approval for a small or major change to a process, piece of equipment, a foundation, etc., from the Owner or customer, which requires changing the respective item's or items' engineering design. This change requires an engineering review by the EPC's and customer's or Owner's engineers because the affected AFC drawing or drawings has to be changed. The onsite work affected by this FDR must also be stopped until this newly revised and approved AFC has been issued by the EPC Company for construction.

FOB (Free-On-Board) – This is the case in which the supplier pays for all shipping charges for the equipment from the factory to the port or city that the customer or Owner specifies in the contract. If the equipment arrives damaged, then the supplier must deal with the insurance company and not the customer or Owner.

I&C (Instrumentation and Control) – This is a specific engineering discipline in which the focus is on the design, installation, and operation of all of the project's instrumentation and the control system for various mechanical and electrical systems.

ITP (Inspection and Test Plan) – This is a factory quality plan required by the project's contract for all equipment or material that will be permanently installed onsite and must be approved by the Owner or customer after each EPC Company's supplier has completed its compilation. The EPC's equipment or material supplier must invite the customer to perform an inspection when there is a "Hold" or "Witness" point for them in the ITP.

KPI (Key Performance Indicators) – These KPIs are specific goals that an EPC Company must meet over the life of the project as per the contract, and failure to do so usually leads to either a penalty of some type or liquidated damages that have to be paid to the Owner or customer. The one standard KPI is in the area of Health, Safety, Environmental (HSE), which is zero Lost Time Incidents (LTIs), where a worker is either severely injured and out of work for a long time or dies at work on the construction site. A very effective and well-communicated HSE Plan is critical to ensuring that this KPI is met over the life of the project.

MAR (Material Approval Requests) – This document is a formal document that is sent by the EPC Company to the Owner or customer for the usage or procurement of material that is not specified in the contractor. It usually involves a review by the Owner's or customer's site Engineering team before it can be approved or not approved for use or procurement.

MDL (Master Drawing List) – This document contains all of the construction drawings required to produce the project's specified deliverables, in the form of a list broken down by engineering discipline, along with each drawing's status. The MDL is share by both the EPC Company and the Owner or customer. If a respective WBS is added for each drawing, it can become a great tool for internal project alignment.

MOM (Minutes of Meeting) – This document is a written recording of a meeting, which must contain the following information:

- All attendees at the meeting
- A specific agenda
- The items discussed, with a breakdown of which item is open or closed
- The person or company responsible to close each open item and the date required for this closure
- A specific date and location for the next meeting

After this meeting is concluded, the company or person who chaired the meeting must compile the MOM and ensure that the specified managers from each company who attended this meeting sign it before it is officially distributed throughout the project organization.

NCR (Nonconformance Report) – This document is presented to either the EPC Company or one of its subcontractors by the Owner's or Customer's QA/QC Department when a specific quality standard has been violated. The number of NCRs over the life of the project, which should be zero if the EPC Company maintains rigid quality control over the life of the project, is also another KPI that is included in the project's contract,.

NOI or RFI (Notice of Inspection or Request for Inspection) – This document is provided to the Owner's or customer's QA/QC inspectors or engineers to inspect a particular work activity and approve it before the subcontractor can move to the next activity.

P&ID (Piping & Instrumentation Diagram) – This is a mechanical drawing that not only shows all of a system's components, such as piping, valves, instrumentation, etc., but also what other systems it interfaces with during operation. The instrumentation typically includes pressure switches, thermocouples, flow switches, thermometers, etc.

TASK		
Civil Foundation Excavation		
BT	WT	ET
45	65	52

PERT (Project Evaluation Review Technique) – This is a scheduling technique that is similar to the Critical Path Method (CPM) in that it attempts to establish one timeline and schedule the work for the project along that line. However, they differ in how an activity's duration is determined. A **PERT** Task would look like the following before its calculated duration is completed:

The BT is the Optimistic Time (**B**est **T**ime), WT is the Pessimistic Time (**W**orst **T**ime), and ET is the Estimated Time (**E**stimated **T**ime), which could be based upon historical data and is considered to be more realistic. Since there may be human error, an activity's duration for PERT is calculated using a weighted average (WA) expressed in the formula, WA = (BT+4xET+WT)/6. The Task shown above would have a WA of 70 days.

PLP (Project Logistics Plan) – This document is compiled by the EPC Company's Procurement Department that details not only the scheduled arrival of all equipment and material (which is provided by the vendors and manufacturers) but also the port of inspection and what customs clearance documents will be required to effect payment by the Owner or customer. This logistics plan will also provide the name of the shipping and hauling companies that will be bringing the equipment and material from the factory of origin to the port, as determined in the contract, as well as from this port to the construction site. If the project is in the United States, this will involve receiving state government permits for certain heavy loads and the specific roads that will be used to bring this equipment and material to the construction site.

PMS (Project Management System) – This is a series of components that, as explained in Chapter 2 of this book, when put together correctly, produce a system that is used by Project or Program Managers to manage a project.

PRMP (Project Risk Management Plan) – This is a document that, as discussed in the Chapter 2, not only lists all of the risks, both internal and external to the project, but also evaluates each one in terms of importance and impact on the project. After this evaluation is completed, the next step is to provide a mitigation plan for each one of the risks identified in the PRMP.

PTW (Permit-to-Work) – This is a permit supplied by the HSE Manager of the EPC Company to a subcontractor stating that this subcontractor is permitted to perform work in this area of the construction site. A subcontractor found performing work without a PTW can be removed from the site.

SS (Site Survey) – This is the initial survey of the complete construction site by a professional survey team that establishes the vertical, horizontal, and lateral benchmarks for each area where a building or some type of structure will be erected on the construction site. This also includes establishing the location of all site access roads and drainage areas for the construction site.

WBS (Work Breakdown Structure) – This is the result of taking one task and breaking it down into many subtasks. For example, if we first take all of the work activities to build just one building and then develop a set of subtasks for each of these activities, then the complete set of all these activities and their respective subtasks is the WBS for this building, which we can then develop into a schedule for that building to be built.

Index

(continued on following page)

(continued on following page)

Printed and bound by CPI Group (UK) Ltd, Croydon, CR0 4YY

17/10/2024

01775655-0013